VICTORIAN
ENGINEERING

VICTORIAN
ENGINEERING

L.T.C. ROLT

With thanks to Mike Chrimes and Annette
Ruehlmann at ICE for the illustrations

First published by Allen Lane/ Penguin 1970
This edition published 2010
Reprinted 2013

The History Press
The Mill, Brimscombe Port
Stroud, Gloucestershire, GL5 2QG
www.thehistorypress.co.uk

British Library Cataloguing in Publication Data.
A catalogue record for this book is available from the British Library.

ISBN 978 0 7509 4657 5

Typesetting and origination by The History Press
Printed in Great Britain

Contents

Foreword

In less than a century, Britain was transformed from a rural economy into the most powerful industrial nation on earth. Until the opening years of Victoria's reign, the pace of life was still largely governed by the motive power of the horse. Communication of information and carriage of goods were slow, while bad roads and cumbersome horse-drawn wagons meant that travel over any distance was limited for all but the very wealthy. By the time Victoria came to the throne at the end of the 1830s, it was clear that momentous events were changing this ancient pattern, for the railway network was under way, cutting through historic landscapes and towns and carrying all before it. It demanded unprecedented civil engineering works, bridges of undreamed-of spans and the formation of a permanent way, all to provide a level road for the steam locomotive, which emerged as a clumsy, clanking monster from the mines to become the sleek and powerful symbol of the age.

While the railway may be the most obvious symbol, it is actually only one part of an infinitely multi-layered story, in which advances in the stationary steam engine as industry's prime mover and the development of increasingly sophisticated specialist tools for the manufacture of machines were fed by and in their turn fed into the advance of the revolutionary infrastructure. Underpinning these advances in transport, manufacture, civil, mechanical, marine and agricultural engineering was iron (and later steel), the production of which increased hugely as the century wore on. Much of the iron came from the Black Country, 'the fiery fountainhead of the Industrial Revolution', which provided the components for everything from steam engines, boilers and rails to splendid arched trainshed roofs and the gigantic bridges and ships made from rolled wrought-iron plates. Slotting into this complex story is the invention of the electric telegraph, which revolutionized communication of information, and the immense improvements in living standards and in public health brought about by the development of gas, and later electric, lighting and crucially by the provision of drainage and an unpolluted water supply. All this was the achievement of a group of remarkable men, the engineers, whose unfettered imagination allowed them to face and overcome every challenge.

L.T.C. Rolt draws all this together into a broad and compelling narrative sweep, taking the reader through the heady days of the 1840s up to the triumph of the Great Exhibition in 1851 down to the end of the century and the 'sorry sequel' of the introduction of the internal combustion engine and the motor car, alas due not to British but to French and German engineers. The celebrity names of Brunel and Stephenson are here but here also are the extraordinary achievements of a host of other, no less brilliant men, whose names have faded into obscurity. Rolt shows why it was that Britain was uniquely placed to engender such an explosion of genius but also reveals the reasons for her failure to retain her technological lead and her consequent eclipse by the European nations and the United States at the end of Victoria's long reign. *Victorian Engineering*, written over 30 years ago, still stands as the best and most readable survey of a period that continues to fascinate and enthrall us.

Julia Elton, President, The Newcomen Society for the History of Engineering and Technology.

Introduction

Because one old widowed lady at Windsor chanced to live so long and because, in the course of her reign, the achievements of British engineers brought about the greatest social revolution that the world had ever seen, the title of this book covers a vast field. To call it a dense wood rather than a field would be a more useful analogy. From the stand-point of engineering history, the subjects I have chosen for my ten chapters could each be readily expanded into as many books. So I make no apology for keeping my eye on the whole wood rather than on any individual trees within it.

I have concentrated only on what I regard as the broad lines of the most significant engineering developments of the period and, with the layman in mind, I have tried to avoid a lot of technical detail. Instead, I have been concerned to emphasize the social consequences of these developments and how they influenced, and were in turn influenced by the thought and opinions then current. To do this has meant leaving the realm of historical fact for one of speculation in which an author's personal opinions and beliefs necessary play a large part in the interpretation of history. I make no apology for this either.

We do no glean wisdom from historical facts but only from our interpretation of their significance. That is why history that consists of a mere recital of facts makes such dull reading. It is the conclusions which the historian draws from these facts that we find stimulating even though we may disagree violently with his point of view. If we are to profit by the lessons of history, the historian has a positive duty to stimulate us in this way. If he fails to do this he is wasting his time and 'history is bunk'. In this connexion it may be significant that the famous American who coined this phrase realized the importance of the work of our pioneer engineers long before we did and carried many examples away to his own Ford Museum at Dearborn.

For the benefit of those who find the engineering history in this book too superficial and wish to pursue any particular branch of the subject in greater detail, I have added a Select Bibliography as a guide to further reading. Many of the titles listed in this are still in print and those which are not should be readily obtainable from a good reference library. They were freely consulted in the preparation of this book and I would like to acknowledge my indebtedness to the authors concerned.

Victoria's reign falls into two distinct epochs. In the first half of the nineteenth century, conditions in the new industrial areas were often appalling and there was considerable unrest. This was the period of the Chartist riots, of the 'Hungry Forties' and of Peterloo. And yet, in spite of all this, the prevailing mood was one of optimism, an optimism of which the Great Exhibition of 1851 was the ultimate expression. Despite all the hardships there was a feeling that these were but temporary growing pains. Even its most oppressed and dedicated opponents felt that by united effort industrial capitalism could be speedily changed into some more equitable system by which Britain's material progress would lead the world into a new era of peace and prosperity for all.

It is a paradox that in the second half of the century, outwardly so peaceable and prosperous that we look back upon it as to some golden age, this mood of optimism, this sense of purpose and direction was lost. Although we are apt to regard these late Victorians as complacent and self-satisfied, theirs was in fact an era of disillusionment and doubt when many of the more acute and perceptive minds were deeply disturbed and bewildered. As I have tried to show, because the Victorian engineers formed the advance guard of the Industrial Revolution, their activities were a major cause of this change of mood and they were, in turn, profoundly affected by it.

In order to trace the sequence of engineering developments in many different fields, this book does not follow a strictly chronological order. Nevertheless it does, I hope, reflect this paradoxical contrast between the two epochs, ascending to, and descending from, the central peak of the Great Exhibition. For it was in that glittering secular temple of iron and glass that the achievements of the Victorian engineer were uniquely celebrated. It was his finest hour.

L.T.C.R.

ONE

The Railway Engineers

The date was Monday, 13 June 1842, and the place Slough station. A train stood at the platform headed by *Phlegethon*, one of the latest seven-foot express locomotives of the Great Western Railway. A group of black-coated officials paced the platform, nervously consulting their watches from time to time, their tall hats gleaming scarcely less brightly in the sunshine than the glossy green paint and polished metalwork of the locomotive. Mr Charles Russell, the Chairman of the Company, was there with the Secretary, Mr Saunders; so were Isambard Kingdom Brunel, the Engineer-in-Chief, and Daniel Gooch, the Superintendent of the Locomotive Department. For the train was no ordinary one and the occasion was a milestone in the history of Britain's railways. The young Queen, accompanied by her consort, was about to make her first railway journey.

Prince Albert had already used the railway several times for journeys between London and Windsor. Foreseeing that it could only be a question of time before that progressive young man persuaded Her Majesty to try the new mode of travelling, the directors had ordered to be built at Swindon a magnificent Royal Saloon. This had first been used in August 1840 by Dowager Queen Adelaide. Now, after many months of anticipation, it was standing at Slough platform, ready for the great event. The Royal carriage was marshalled between two ordinary saloons; an open braked

1

second-class coach was coupled next to the engine and three open carriage trucks brought up the rear.

At a little before noon the cavalcade from Windsor drove into the station yard and Charles Russell ushered the Queen into the 'splendid apartment' prepared for her, while the carriages of her attendants were loaded upon the waiting trucks. This done, she took her seat in the Royal Saloon and, punctually at noon, the train moved off with Brunel and Gooch riding on the footplate of *Phlegethon*.

Some inaccurate press reports of the occasion gave birth to the long-lived legend that Brunel actually drove the engine. But, in giving evidence before a Parliamentary Committee in the following year, the great engineer admitted with disarming frankness: 'I never dare drive an engine . . . because if I go upon a bit of the line without anything to attract my attention I begin thinking of something else.' So it may have been as well for England and for her railways that it was Daniel Gooch and not Brunel who held the regulator on this historic occasion.

At the London end of the line, the most elaborate arrangements had been made. Says a contemporary press report:

> At Paddington by 11 o'clock the centre of the wide space apportioned for the arrival of trains was parted off and carpeted with a crimson carpet, which reached from one end of the platform to the other. The whole of the arrangements for the reception of the Royal party were under the superintendence of Mr Seymour Clarke, the Superintendent of the line, assisted by Supt. Collard of the Company's Police. . . .
>
> Before 12 o'clock large numbers of elegantly dressed ladies consisting of the families and friends of the Directors and Officers of the Company were arranged on each side of the part apportioned for the arrival of the Royal train, and at five minutes before 12 o'clock

Her Majesty's Carriage drawn by four horses arrived from the Royal Mews at Pimlico, and a few minutes afterwards a detachment of the 8th Royal Irish Hussars under the command of Capt. Sir G. Brown arrived from the Barracks at Kensington for the purpose of acting as an escort to Her Majesty.

Precisely at 25 minutes past 12 o'clock the Royal Special train entered the Terminus having performed the distance in 25 minutes, and on Her Majesty alighting she was received with the most deafening demonstrations of loyalty and affection we have ever experienced. . . .

Over the centuries London had welcomed many a king and queen, but this occasion was unique. Never before had a reigning sovereign made such an entry into the capital city. No one of those who cheered the young Queen could possibly have realized how fitting the occasion was; for Victoria's long reign would give a name to an age, an age to be forever preeminently associated with steam power and steam railways.

Railways, in the modern sense of the term, had been born in 1830 with the opening of the Liverpool & Manchester line. The first main lines, London & Birmingham, Grand Junction, North Midland, Great Western, London & Southampton, had already, by 1842, formed the first surprisingly healthy growth of a national railway system. It was surprising because a people who, since time out of mind, had regarded the horse as the only proper means of overland locomotion were not to be lightly won over to the new steam monster. This was not a simple matter of conservative prejudice. A great pre-Victorian generation of engineers had given Britain a network of metalled roads and canals which, at that time, represented a prodigious and unique capital investment in transport. By enabling the horse's potential to be used to the full, either in drawing passengers and mails down the new roads or in hauling boats and barges heavy-laden with minerals and merchandise on the canals, this investment had been well justified. A huge and flourishing coaching industry, the rich dividends of the canal companies, the rapid expansion of trade and manufacture, all proved to the public how well it had paid off. Were such profitable assets to be cast aside as worthless for the sake of a new-fangled steam toy?

This was the question so eloquently put to Parliament and people by the powerful vested interests involved, by Turnpike Trusts, by coach proprietors, horse-dealers and canal companies. They played upon the natural conservatism of the public by exaggerating the dangers of steam travel. Passengers would be mangled under the wheels of the fiery juggernaut if they were not suffocated in tunnels or by the sheer unnatural speed of their passage through the air. Before its hot breath, crops would wither and animals sicken and die. They also pointed out that whereas all were free to use the roads or canals on payment of tolls, the villainous new railway companies would hold a monopoly of the traffic on the iron roads which they proposed to build across England.

In the face of such arguments, the steam railway might never have been born in 1830 had it not been for the almost fanciful faith of George Stephenson allied with the capital provided by a minority of merchants and manufacturers. These men alone realized that roads, and more particularly canals, had been doomed by their own success; that the expansion of commerce they had helped to generate had outgrown them. But amongst the majority of Englishmen the prejudice against railways was still strong. It was the Queen who finally overcame this prejudice when she made her historic journey from Slough to Paddington and so set the seal of Royal approval on the railways. So, although the seeds were sown and had sprouted before the Queen came to the throne, it was during her reign that railways came to their mature and prosperous flowering. Nor did this golden age of railways long outlive her. The idea of the railway as the supreme material achievement of the Victorian age may not be historically exact but it remains true none the less.

The first canals, built cheaply in the 1760s, earned such fantastic dividends for their proprietors that they induced a 'canal mania' of promotion and speculation in the period between 1790 and 1794. The canal companies that were then authorized had to build their lines in a period of rapid inflation – caused partly by the French wars and partly by the speculation itself – and as a result they laboured under an immense burden of capital debt and never enjoyed the same financial success as their predecessors. Now, despite the fact that these events had occurred within the memory of persons still living, history repeated itself with depressing exactitude. The success of the first main lines and the seal of Royal approval swayed public opinion from hostility and scepticism to an opposite pole of enthusiasm and gullibility. This culminated in the 'Railway Mania' of 1845–6 which, brief though it was, made the previous canal boom pale into insignificance.

The serio-comic story of the Railway Mania has often been told. Here it is only relevant to note its effects. The efforts of Gladstone's Ministry to stem the tide of new Railway Bills were swept aside and by November 1845 623 new railway schemes had been laid before Parliament representing a capital outlay of £563 million. Although only a fraction of these projects were actually authorized and carried out, the results were startling enough. By 1845 a sum exceeding £67 million had already been spent on a little over 2,000 route miles of railway. In that year alone 2,170 route miles of new line were authorized and, since Brunel estimated that construction costs had risen by fifty per cent, the capital investment cannot have been less than £100 million. But the bubble soon burst and by 1848 the mania had become a matter of history.

The end of George Stephenson's active career virtually coincided with Victoria's accession though he lived on in retirement until 1949. Prime author of the railway revolution, its headlong pace astounded and appalled him. In 1830, the fact that there were then eight Railway Bills before Parliament had moved him to write: 'It is really shameful the way the country is going to be cut up by Railways. . . .' So what can he have thought of the events of 1845? It was those railway engineers on whom the old man's mantle had fallen who now had to bear the burden and heat of the day, and they regarded the speculative mania with equal distaste despite the work it brought them. George's only son, Robert Stephenson, repeatedly inveighed against the profligate waste of capital on competing railway schemes, while George's ex-pupil, Joseph Locke, from his experience of railway-building in France, advocated direct Government control. But it was their rival, Isambard Brunel, who, in his usual trenchant style, best expressed the engineer's view of the situation in a letter to a friend in France:

> Here the whole world is railway mad [he wrote], I am really sick of hearing proposals made. I wish it were at an end. I prefer engineering very much to projecting, of which I keep as clear as I can . . . The dreadful scramble in which I am obliged to get through my business is by no means a good sample of the way in which work ought to be done . . . I wish I could suggest a plan that would greatly diminish the number of projects; it would suit my interests and those of my clients perfectly if all railways were stopped for several years to come.

But it was not to be. This great triumvirate of railway engineers was too caught up in the revolution they had helped to engender for there to be any possibility of withdrawal. Their finest work was still before them. Stephenson, Brunel and Locke, and other scarcely less eminent railway engineers like Charles Vignoles and William Cubitt, scrupulously avoided associating their names with any of the grandiose 'bubble' schemes put forward to attract the speculator, although there was considerable financial temptation to do so. Any one of these famous names on a company prospectus was worth a great deal to unprincipled promoters, and those lesser mortals in the profession who fell for the promoters' bait were looked upon with contempt by such men as Stephenson, Brunel and Locke. They would only undertake projects which they judged to be of lasting benefit and, in doing so, far from merely selling their names, they shouldered a degree of personal responsibility inconceivable in later, more highly organized days when the engineering profession had become more specialized.

The extension of rail communications between the capital and Scotland, Ireland, South Wales and the furthest west of England became the chief concern of the great triumvirate in the period of the railway mania. The rails had already reached from the Thames to the Tyne at Gateshead and Robert Stephenson was engineer of the Newcastle & Berwick, authorized in 1845 and destined to become a link in the east-coast route to Scotland. Meanwhile, Joseph Locke had begun, in 1844, his extension of the rival west-coast route to the north from Lancaster to Carlisle. Robert Stephenson was also appointed engineer of the Chester & Holyhead Railway which was likewise authorized in 1845, while in the south Brunel was extending his broad-gauge lines westwards from Gloucester along the seaboard of South Wales to the port of Milford Haven and from Exeter towards Plymouth and Penzance.

From the point of view of civil engineering construction, such major lines of railway differed little except in magnitude from the earlier canals and the railway-builders benefited greatly from the past experience of their predecessors. Nor was that experience by any means confined to the top of the railway-building pyramid as the name 'navvy' (navigator) makes clear. Nor had methods changed appreciably in the interval. In place of the horse tramways used by the canal-builders, steam locomotives on temporary railways might be used, but this apart there was still no substitute in Victorian England for the muscles of men and horses or the power of black powder to blast the rock in tunnel or cutting.

But in one respect the railway age compares very ill with the canal age that went before it. Whereas the long list of canal and road works for which Thomas Telford was responsible had been completed without a single serious accident, the railways were built at a terrible cost in human life. The reason for this was the rapid growth of commercialism which brought increasingly heavy pressure to bear on all ranks from superintending engineer to navvy, forcing them to take risks in order to speed the day when a new line would begin to earn revenue. Telford, towards the end of his life, had already marked this tendency and had protested against 'such haste pregnant as it was, and ever will be with risks'. But in a new and brash profiteering world, the old engineer's warning went unheeded. What we now refer to as the commercial rat-race had already begun in early Victcorian England, but because it had not yet produced ingenious machines to replace men, it took its toll in flesh and blood.

It is strange that some of the techniques perfected by the canal engineers appear to have been forgotten and worked out afresh by their successors. For example, his contemporaries credited Robert Stephenson with originating the construction of skew arches with winding masonry courses on the London & Birmingham Railway, whereas visual evidence survives to this day to show that the canal engineers pioneered this form of construction.

Nevertheless, in bridge-building particularly, the railway engineers had to solve some formidable problems such as had never confronted the builders of roads or canals. Telford had been the first master of iron in civil engineering. He had built slender arch bridges in cast iron and used wrought-iron chains to sling his Holyhead road over the Menai Strait. But neither the cast-iron arch nor the suspension bridge was suitable to bear the great weight of a steam railway train. Some new technique had to be devised.

The earlier railways were laid through comparatively easy terrain and it was not until building was extended through the more difficult country of the north and west in the 1840s that the problems of spanning Tyne and Tweed, Conway and Menai, Wye and Tamar called forth all the engineering greatness of Stephenson and Brunel.

As a material for railway structures, cast iron is brittle and treacherous. A cast-iron beam posssesses immense strength in compression but is fatally weak in tension. The railway engineers knew this. For his great high-level bridge over the Tyne between Gateshead and Newcastle, the first step on the new road to the Scottish border, Robert Stephenson designed bow-and-string girders of cast iron resting on five stone piers at a height of 146 feet above the river. This bow-and-string design is one calculated to avoid excessive tension in the members, but even so, Stephenson was taking no chances. The spans are short, the members massive and particular care was taken over their casting and testing. The Newcastle high level bridge, begun in 1846 and completed in three years, stands to this day, still carrying road and rail traffic on its two decks, the last and greatest monument of cast-iron bridge construction. The Royal Border bridge over the Tweed at Berwick, a towering stone viaduct of twenty-eight spans, was completed a year after the bridge at Newcastle. Both were formally opened by the Queen. On the second occasion, in a train drawn by a locomotive extravagantly bedizened in a livery of Royal Stuart tartan, she was able to travel direct to her palace of Holyrood.

9

For railway bridges of more modest proportions it had become customary to use simple cast-iron beams trussed with rods of wrought iron to relieve the tension stresses in the lower flange of the beam. The principle was sound, being the same as that employed in the modern reinforced or pre-stressed concrete beam. But unfortunately the method of attaching the rods to the beam was unsound and in 1847, while the Newcastle High Level bridge was being built, such a trussed girder in a bridge designed by Stephenson to carry the Chester & Holyhead line over the Dee at Chester broke under the weight of a locomotive and plunged its train into the river. This disaster, and the inquest which followed, caused a furore which led to the appointment of a Royal Commission to inquire into the use of iron in railway structures. As a result, the use of cast iron for bridges of large span fell into disfavour except in columns which were subject only to compressive stresses. For all beams, wrought iron took its place. That this change came about so swiftly was largely due to the lessons learnt in the design and construction by Robert Stephenson of the Britannia Bridge over the Menai Strait. Whether this bridge is judged by its advance on what had gone before, or by its influence on future engineering practice, it must undoubtedly rank as the greatest and boldest civil engineering feat of the early Victorian era.

When George Stephenson had originally surveyed a line from Chester to Holyhead in 1840 he had proposed to utilize one carriageway of Telford's suspension bridge for the crossing of the Menai Strait, drawing the trains, divided if necessary, over the bridge with horses. The authorities responsible for the Holyhead road, however, would only permit this as a temporary expedient, so the idea was abandoned. The Act authorizing the railway was passed in 1845 – most unusually – with a gap of five miles in the plans, the problem of the Menai rail crossing being still unsolved.

Robert Stephenson finally settled upon a site for the crossing at a point a mile south of Telford's bridge where the Britannia Rock in mid-channel offered a convenient base for a central pier. Here he at first proposed to construct a cast-iron bridge of two great arched spans springing from a massive central pier and abutments of masonry. This design was turned down by the Admiralty who, in the interests of shipping passing through the Strait, insisted on limiting the size of the central pier and on a uniform headway of 105 feet throughout between pier and abutment which ruled out any form of arched construction. This faced Stephenson with the problem of devising some form of flat girder bridge. When it is realized that this involved two spans of 460 feet each, whereas the largest girder bridge so far built had a span of only sixty feet, the magnitude of the problem will be appreciated.

Some form of rigid tubular or box girder, supported by chains like a suspension bridge was at first envisaged. But whereas in a true suspension bridge both deck and chains are flexible, the deck falling or rising as the chains expand or contract, there were serious objections to the combination of flexible chains and rigid beams. So Stephenson's mind turned towards some form of girder construction that would be self-sustaining without the aid of chains. Influential opinion, however, represented by General Pasley, the Chief Inspecting Officer of Railways, and Professor Eaton Hodgkinson, the foremost authority on the theory of iron beams, held that the use of chains would be essential. Throughout the subsequent experiments on which the final design was based, Stephenson wisely side-stepped controversy by concealing his intentions, stating that since chains would be in use in the construction of the bridge,* it would be logical to retain them afterwards to give additional support.

---

* At this time he envisaged slinging a temporary deck from the chains, over which the completed tubular girders would be run into position over temporary tracks, an equivalent weight being withdrawn at the same time.

Although there was a growing body of practical experience in the use of wrought iron in boiler-making and ship-building, there was no theoretical knowledge whatever of its properties when used as a beam at the time Robert Stephenson, assisted by Eaton Hodgkinson and William Fairbairn, the ship-builder, embarked on their classic series of experiments at Fairbairn's shipyard at Millwall to determine the best form of girder for the Britannia Bridge. This may seem surprising in that the future of railways depended as much upon the successful production of the rolled wrought-iron rail (itself a simple form of I-beam spanning the sleepers) as upon the steam locomotive.

In a classic series of experiments in which wrought-iron beams of circular, oval and rectangular section were loaded at the mid-point until failure occurred, this team of three not only achieved their immediate purpose but laid the foundations of modern structural engineering theory. The form of girder they decided upon for the Britannia Bridge was of hollow rectangular section, large enough for a train to pass through it. The top and bottom were to be of double plate of cellular construction and the sides of single plate stiffened with vertical angle irons to prevent buckling. Because, in the final experiment, a one-sixth scale model of such a tube withstood a weight of eighty-six tons before failing, nothing more was heard about supporting chains. Nor were they used in construction, it having been decided to build the main tube sections on stages along the Caernarvon shore of the strait, float them out to the site and then lift them into position on the tall stone towers by hydraulic means, building up the masonry beneath them. As the bridge was to be for a double line of railway, there were four main tubes. Stephenson finally resolved to unite each pair of these main tube sections *in situ* not only with each other but with the two half-sections of tubes built out from the abutments to the land towers. There would thus be formed two continuous tubes, each 12,511 feet long, anchored on the central Britannia tower and extending from abutment to abutment. This decision to unite the tubes into one continuous beam contributed greatly to the strength of the structure.

To roll large plates or girders of I-section was beyond the capacity of the contemporary iron industry so that the tubes, each main section of which weighed over 1,500 tons, were made entirely from small plates and angle sections riveted together. Their movement controlled by cables and manually operated capstans, these great tubes were successfully floated into position on pontoons.

Identical tubes, each 400 feet long, were used to span the estuary of the Conway and these, too, were floated into position. This small bridge was completed in 1848 and afforded valuable experience for the greater task that lay ahead at the Menai. There, the first main tube was successfully floated and raised in June 1849, the bridge being ready for single-line traffic in the following January. On 18 March 1850 the bridge was completed and opened for public traffic. The iron way was open to Holyhead.

It speaks volumes for the sterling qualities of Stephenson and Brunel that, while their professional rivalry was intense, their engineering policies being poles apart, they were yet able to maintain a close personal friendship. 'It is very delightful,' wrote Brunel, 'in the midst of our incessant personal professional contests, carried to the extreme limit of fair opposition, to meet him [Stephenson] on a perfectly friendly footing and discuss engineering points.' So, when Stephenson was about to float out the first great tube of his Britannia Bridge, Brunel dropped all his own engagements and hurried to North Wales to be at his friend's side and give him moral support.

Two years later, Brunel found himself standing on the bank of the Tamar at Saltash Ferry facing an almost identical problem. The tidal Tamar, or Hamoaze, at Saltash is 1,100 feet wide and here again the Admiralty stipulated a headway of not less than 100 feet and that the channel should not be obstructed by more than one pier. In one respect the problem was much more difficult, for here there was no convenient half-tide rock in mid-channel to provide a base for this single pier. Soundings showed that the essential rock foundation lay beneath eighty feet of water and mud. The founding of a pier at this depth was no less a pioneer achievement than were Stephenson's tubes at the Menai.

An iron tune, eight feet in diameter and eighty-five feet long, was first sunk down to the rock. Within it, trial borings were made into the rock and its profile plotted. Then a 'Great Cylinder' thirty-five feet in diameter was constructed on Tamar shore, one end being shaped to suit the rock profile. This was towed out, up-ended and sunk on the site of the pier.

Brunel designed this cylinder to be part pressure caisson and part diving bell. A series of pressurized chambers round the periphery would be used for the excavation necessary to sink it down to the rock and then to exclude water from the central portion within which the pier would be built. In this way he hoped to avoid having to build the pier under pressure which would mean passing both the men and the materials through an air-lock chamber. In the event, however, this intention was defeated by an unsuspected fissure in the rock. So far from excluding water, the pressure in the outer compartments forced it through this fissure into the central portion in a volume that overpowered the pumps. Consequently the whole of the great cylinder had to be modified and strengthened *in situ* to form one large pressure caisson, the largest that had been employed in civil engineering at that time. This extremely difficult and delicate operation was carried out successfully and the pier built. In these modern days of concrete construction, such a caisson usually forms the shuttering for the concrete and remains as part of the permanent structure, but Brunel's cylinder was designed in two halves and when the masonry pier had been built it was removed.

The reason why Stephenson selected a rectangular tube for his Menai Bridge was that when he began his historic series of experiments by testing tubular beams of round and oval section, so little was known of the properties of wrought iron that the thickness of metal used in trial tubes was quite inadequate. Consequently the upper surface of the tubes buckled under compression when they were loaded at the mid-point and Stephenson decided to abandon girders of this form. Brunel, on the other hand, had great faith in the strength of iron tubes which he confirmed in a similar series of practical experiments, and he subsequently used them in a bridge to carry the line to South Wales over the Wye at Chepstow. In this the double-line deck of the bridge was suspended by chains and rods from two wrought-iron tubes, each nine feet in diameter and 300 feet long, carried fifty feet above the deck on masonry towers, arching over the tracks. This Chepstow bridge proved so successful (it was only rebuilt in 1962) that Brunel resolved to use a similar design for his great bridge at Saltash. Here the two main spans required were almost identical with those at the Menai, 465 feet each, but for economic reasons it was decided to build the bridge for a single line of metals only; hence only two tubes would be needed. Brunel designed these in oval arched form, 12 feet 3 inches high by 16 feet 9 inches broad in section, to support the bridge deck below.

As Brunel has often been charged with extravagance, it is worth pointing out that this was a more economical form of construction than the rectangular tubes used at the Menai. Like the latter, each main span of deck and tube was built complete on the shore of the Tamar, floated out into position and then raised by hydraulic power. Owing to the financial difficulties of the Cornwall Railway Company, construction was protracted. It was not until 1859, the year of Brunel's death, that the Saltash bridge was completed and formally opened by Prince Albert who bestowed his name upon it.

These two great bridges stand to this day, the Britannia with its huge stone couchant lions and the Royal Albert with the simple inscription 'I.K. Brunel, Engineer, 1859' high over the tower arches. They are outstanding examples of what has been rightly called the heroic age of railway engineering.

Of the many wrought-iron railway bridges to be built subsequently in Britain before the introduction of steel, three were outstanding. These were, in order of completion, Crumlin Viaduct (1857), Belah Viaduct (1861) and the Severn Bridge (1879). All three have, unfortunately, been demolished in recent years.

Crumlin Viaduct was built to carry the Taff Vale extension to the Newport, Abergavenny & Hereford Railway over Ebbw Vale and, at the time of its completion, it was claimed to be the largest viaduct in the world. Its total length of 1,650 feet was divided into two sections by the crest of an intervening hill. Of these, the main eastelry section carried a double line of rails 200 feet above the river Ebbw in spans of approximately 150 feet. Wrought-iron lattice girders were supported on trestles, each consisting of four cast-iron tubes braced with wrought-iron lattice work. It was designed and built by the contractor T.W. Kennard who established a works in the valley directly beneath where all the components of his great viaduct were fabricated. Crumlin was an outstanding achievement for its date and its success was such that Kennard received orders from abroad for several similar structures, all the parts of which were produced at his Crumlin works.

The viaduct over the Belah valley between Barras and Kirkby Stephen was only narrowly surpassed in size by

Crumlin, being 1,040 feet long and 196 feet high. The form of construction was the same and it was the largest of three similar viaducts built by Thomas Bouch to carry the South Durham & Lancashire Railway across the Pennines. It was doubtless the success of these structures which emboldened Bouch to embark on his ill-fated scheme for a great bridge over the Firth of Tay, of which more later.

The bridge over the estuary of the Severn at Sharpness was built by the Severn Bridge Railway Company with the powerful backing of the Midland Railway, with whose Sharpness Docks branch it would connect, primarily to provide a more direct rail outlet for coal from the Forest of Dean. The Severn at Sharpness is nearly three quarters of a mile wide and the total length of the bridge, including a substantial masonry approach viaduct at its western end, was 1,387 yards with a height above high-tide level of seventy feet. The piers consisted of cast-iron cylinders assembled in sections and filled with concrete. Compressed air was used in sinking these piers through the sand down to bedrock at a maximum depth below high water of seventy feet. On these piers rested twenty-one spans of wrought-iron bow-and-string girders, fourteen of 134 feet span, five of 171 feet and, over the main channel of the river, two of 327 feet. In addition a swinging span of 196 feet weighing 400 tons was required to carry the railway over the Gloucester & Berkeley Ship Canal. Unlike the rest of the bridge, this last was built to accommodate two lines of metals. It revolved on cast-iron rollers, a central masonry pier carrying the roller track. It was swung by steam power, an overhead cabin housing the vertical boilers and horizontal steam engines which engaged the bridge gearing through ribbed friction drums.

Because of the great tidal rise in the estuary and the speed of the current which can run at nine knots when a spring tide is making, the consulting engineer, T.E. Harrison, considered it unwise to attempt to float the girder spans out complete and lift them into position as had been done in the case of the Britannia and Saltash bridges. Instead, all the iron work was assembled *in situ* on timber stagings. Despite this, and the fact that tides overset some of the pier sections before they could be completed and filled, the bridge was completed in less than four years from the time the contracts were let in March, 1875.

The bridge was opened with great pomp and ceremony in October, 1879 when, to the accompaniment of a fusillade of exploding detonators, a distinguished company, including Sir Daniel Gooch and representatives of the government of the United States, France, Germany and the Netherlands crossed the Severn in a special train to be entertained to a banquet at Sharpness. But *sic transit gloria mundi*: the Forest of Dean coal traffic gadually declined until, in a fog on the night of 25 October 1960, a drifting oil tanker barge struck the piers, bringing down two of the smaller spans. Because traffic was insufficient to warrant repair, the great bridge thereafter stood forlornly, derelict awaiting its recent demolition.

Joseph Locke, the third of the great trio of pioneer railway engineers, has left behind him no single memorial as spectacular as the Britannia or Royal Albert bridges. Yet he probably exercised a greater influence on railway engineering in particular and on civil engineering generally than either Stephenson or Brunel. He certainly did more than either to enhance the reputation of British engineering abroad.

Most of Locke's contemporaries grossly underestimated both the cost and the time necessary to complete a new line of railway. Stephenson had estimated that the London & Birmingham would cost £21,736 per mile, whereas the actual cost in round figures was £50,000 per mile. Similarly, Brunel estimated £2½ million for the line from London to Bristol whereas the actual cost was £6½ million or £55,000 per mile. True, these costs were inflated by the unexpected difficulties which both engineers encountered, notably in driving the tunnels at Kilsby and Box. Moreover, we should remember that both these lines were carried out in the early days when engineers may have thought that to allow a sufficient sum for contingencies would simply mean frightening off the promoters so that the line would never be built. But part of the reason why costs were inflated was that these lines were built by a multiplicity of small contractors, the majority of whom were inexperienced and soon went bankrupt. The engineers themselves were partly responsible for this through their failure to furnish the contractors with sufficiently detailed and accurate specifications of the work required of them. Joseph Locke, on the other hand, was the first man to set a new standard of accuracy in estimating by supplying the contractor with carefully detailed specifications and by exercising a closer control over his subsequent performance. He applied these methods to the first major line of railway for which he was responsible, the Grand Junction which linked the original Liverpool & Manchester line with Birmingham. Here the cost per mile was £18,846 as against his estimate of £17,000.

Admittedly, compared with the London to Birmingham or Bristol lines, the Grand Junction was far easier to build, involving few large engineering works, but where some major work was concerned, Locke never shirked contingencies, aiming to give the company a realistic forecast of probable costs no matter how unpalatable. A good example of this occurred when Locke replaced Charles Vignoles as engineer of the Manchester & Sheffield Railway. Vignoles had estimated £98,467 for the great three-mile summit tunnel beneath the Pennines at Woodhead on the false assumption that it would not need to be lined. To the consternation of the company, Locke's first action was to double this estimate. Nevertheless, such was his reputation that the revised figure was accepted.

It was on this bleak Pennine moorland 1,000 feet above sea level and nine miles from the nearest town that the grimmest battle of the railway age was fought. It desplays that age at its best in the sheer pertinacity with which a task of incredible difficulty was tackled, and at its worst in the appalling conditions and in the contempt for human life shown by contractors and resident engineers alike. A force of over a thousand navvies, some with their women and children, lived on the desolate moorland in a shanty town of improvised dry-walled stone huts roofed with heather or ling, sometimes as many as thirteen to a hut. In the tunnel the rock strata were variable and treacherous and streams cascaded from the roof and down the working shafts so that men had to work almost knee-deep in mud and water. When this, the greatest tunnel so far driven in England, was at last completed in December 1845, after five years' work, the number of men killed or maimed by blast, rock falls or other accidents in proporition to the total force engaged exceeded the casualties sustained in any of the major battles of the century, Waterloo not excepted. But the total cost in money was £200,000, almost precisely Joseph Locke's original estimate.

The tunnel was a single-line bore and in 1847 the work of driving a second bore beside it was begun. This proved an easier task but one made equally grim by an outbreak of cholera among the navvies which claimed twenty-eight lives. Finally, the work of driving a new double-line tunnel beside the old ones was begun in 1849. The fact that, with the rock strata known from the original sections and every modern aid, this tunnel took 5½ years to complete is the best tribute to what the railway pioneers could accomplish.

More than any other engineer, Locke was responsible for the rise to power of a new class of men – the great railway contractors. His method of basing estimates on detailed specifications was far more easily applied if he had a few large contractors working for him who understood his requirements. So, with these contractors engaged on Locke's railway works it was a case of the survival of the fittest and a few emerged head-and-shoulders above the rest; men capable of organizing large numbers of that most unmanageable and independent class, the railway navvies, and winning their respect; men to whom an engineer could entrust the construction of an entire railway with complete confidence. Such men were Thomas Brassey, a young Cheshire surveyor, who undertook his first contract on Locke's Grand Junction, and Joseph Firbank, a sub-contractor on the Woodhead Tunnel at the age of twenty-two.

With Brassey especially, Locke built up an extremely fruitful working relationship. In the construction of the London & Southampton Railway he let to Brassey contracts covering 118 miles of line worth £4.3 million and his trust was not misplaced. Brassey matched in energy and ability the great triumvirate and at the height of his power he commanded a labour force of 45,000 navvies on railway works, not only in England but in Europe, most of them under Locke's direction. And though Brassey engaged local labour on his European contracts he thought poorly of it; for the more difficult work he relied upon his *corps d'élite* of English navvies who were then regarded as the finest labour force in the world.

Like the English navvies, British railway engineers enjoyed a supreme reputation in the western world. Robert Stephenson was responsible for railways in Belgium and Sweden, Brunel carried out work in Italy, while Locke, usually with Brassey as his right-hand man, built lines in Holland, France and Spain; from Utrecht to Rotterdam and Amsterdam; from Paris to Rouen and Le Havre and to Nantes, Caen and Cherbourg; from Barcelona to Mataro.

Looking back on this heroic age of railway-building, and having the advantage of hindsight, we can see how these greatest of railway engineers erred according to their individual temperaments. In this respect, Robert Stephenson, although he was sometimes over-cautious and conservative, was probably the soundest of the three. The brilliant

Isambard Brunel was a perfectionist with little concern for the pockets of shareholders. It was this perfectionism which led to his adoption of a broad gauge of seven feet for his Great Western Railway when every other engineer in the country was content with what he contemptuously called 'the coal wagon gauge' of 4 feet 8½ inches which the Stephensons had chosen as their standard. That his was a technical *tour de force* was proved by the overwhelming superiority of broad-gauge performance in the celebrated Battle of the Gauges which enlivened the hectic railway world of the 1840s. This engagement culminated in a pyrrhic victory for the Great Western when that company launched the first service of express trains in the world between Paddington and Exeter, but apart from stimulating technical development on the competing narrow-gauge lines, the broad gauge must be judged a disastrous financial failure. Nevertheless, it was not finally abolished until 1892. Brunel had argued wrongly that the superiority of his broad gauge in smoothness, speed and comfort would be so overwhelming that all other railways would be compelled to convert to it. This was a case of 'Jack's the only boy in step' for it overlooked the considerable route mileage of narrow-gauge lines already in use and the fact that it is much simpler and less costly to narrow the gauge of a railway than it is to widen it. The chaos and inconvenience caused by the break of gauge at points of interchange between the rival systems brought a growing pressure to bear against Brunel's broad gauge. Queen Victoria cannot have been amused when, on journeys between Balmoral and Osborne, the broad gauge compelled her to change trains twice, at Gloucester and Basingstoke.

If Brunel ignored the shareholders, Locke sometimes showed too great a concern for them. Although in the planning and execution of large-scale works he set his profession a new standard of proficiency, his concern to save money led him to choose routes which, while they undoubtedly cut capital cost, saddled the company concerned and their successors with a heavy burden of additional operating expenses.

George Stephenson had maintained that railway routes should be planned and surveyed in the same manner as the earlier canals with the longest possible stretches of level or near level line. Where changes of level became unavoidable, the gradient should be short and steep, trains being assisted by rope haulage if necessary. He based this policy on the sensible axiom that the power of the new steam locomotive should be used most economically in hauling maximum loads at a good average speed and not wasted in climbing gradients. This is still sound theory since the haulage capacity of any locomotive over a given route is always determined by the ruling gradient. In practice, however, owing to the nature of the country, a level route must either be very circuitous or it must involve extremely costly engineering works. One reason for the high cost per mile of the first main lines, the London & Birmingham or the Great Western, was that both Robert Stephenson and Brunel adhered to the theory of the level road.

The steam locomotive developed so rapidly in power and performance, however, that by the 1840s an opposing group of engineers and surveyors had emerged who are usually referred to by railway historians as the 'up and over' school. Locke was the first and foremost exponent of this school. His confidence in the powers of the steam locomotive naturally became popular during the years of the railway mania when so many new lines were promoted, often through difficult country.

The route of the west-coast line to Scotland between Lancaster and Carlisle was the subject of the first and most notable clash between these two schools of thought. The Stephensons roundly declared that a direct line was out of the question and proposed an impossibly circuitous route round the Cumberland coast. Against this, Locke advocated a direct line over Shap Fell that was thirty miles shorter, arguing that the extra cost of the motive power required would be less than the interest on the additional capital needed to build the longer Stephenson line. This argument prevailed and, with his usual efficiency, Locke proceeded to build the railway over Shap. The first sod was cut near Birkbeck in July 1844, and, with as many as 10,000 navvies at work, the line was opened to Carlisle in December 1846. But this speedy completion of the first railway to reach the Scottish border was achieved at the price of a gruelling gradient facing north-bound trains culminating in four miles at 1 in 75 from Tebay to Shap summit which has bedevilled traffic operation from that day to this.

That the Stephenson alternative was not practical is obvious, but the fact that there was a happy medium and that Locke carried his 'up and over' policy too far in his concern for economy was proved as early as 1862. In that year a deviation was surveyed which showed that, at the expense of a mile-long tunnel, the final killing climb to Shap summit could be avoided. This deviation was never built.

North of the border, history immediately repeated itself on the line from Carlisle to Glasgow and Edinburgh through the border hills. A circuitous route via Dumfries and Nithsdale was rejected in favour of Locke's route through Lockerbie, Annandale and Clydesdale which was twenty-five miles shorter but involved a climb of no less than 9¾ miles at 1 in 75 to a 1,000-feet summit between Beattock and Elvanfoot.

To sum up, we can generalize by saying that Joseph Locke's theory prevailed on most of the lines built after 1845. Its disadvantages did not become fully apparent until towards the end of the century when the passengers' demand for improved standards of comfort led to a great increase in passenger train weight. Moreover, such heavy trains had to be hauled at higher speeds to satisfy the public. Today the wheel has come full circle. It is no coincidence that, while so many of the later railways are withering away, it is Robert Stephenson's original London & Birmingham main line which has been selected for electrification and modernization because it is best suited to the fast running which is essential if railways are to be competitive in the modern world.

The canal engineers had acted as their own architects, but when the Stockton & Darlington Railway was being built, the company engaged a Durham architect, Ignatius Bonomi, to assist George Stephenson with the design of the Skerne bridge at Darlington, thus establishing a precedent which persisted throughout the high Victorian age of railway-building. This division of labour between engineer and architect may have given rise to that regrettable Victorian conception of architecture as a form of decorative treatment which could be applied to a building like icing to a cake. The facility with which this was achieved accounts for the proliferation of different architectural styles that marked the period.

Among the great railway engineers, only Brunel insisted upon retaining personal responsibility for architectural design. Only in the design of the present Paddington station, built in 1852–4 to replace the original terminus at Bishop's Road, did Brunel collaborate with Digby Wyatt, but in this case Brunel engaged Wyatt on his own initiative as his personal assistant. But Brunel was an exception that proves the rule. Almost invariably the railway company engaged an architect, not only to be responsible for stations and other buildings, but to collaborate with the engineer in the design of bridges, viaducts or tunnel mouths. This collaboration sometimes makes attribution difficult; it is impossible to determine with certainty where the engineer stopped and the architect began. For example, Francis Thompson, one of the most gifted of the earlier generation of railway architects, was employed by the Chester & Holyhead Railway company and, because he is said to have a hand in the design of the Britannia Bridge, we must assume that Thompson, and not Stephenson, was responsible for the Egyptian style of the portals and the tops of the pier towers.

It was the function of the architect to make the new railways seem respectable and therefore acceptable to the public, or rather to the influential moneyed section of it which mattered. Would the Queen have made her first railway journey when she did had that 'splendid apartment' not been provided at Slough station to receive her? And when the railway eventually reached Windsor she was even more elaborately catered for by Sir William Tite's Gothic essay in brick and stone facings.

The first of the great main lines to enter London celebrated its arrival with Philip Hardwick's massive Doric portico at Euston. Originally forming the central feature of a screen in the manner of Sir John Soane, Hardwick's portico established a monumental convention for the large urban station that was followed throughout the subsequent Victorian era although the results never equalled the magnificent prototype which our own uncouth generation has now destroyed. In the relationship of buildings to platforms and tracks, however, it was not Euston which set the pattern but the original London & Southampton railway terminus at Nine Elms (1838) with buildings by Sir William Tite. Because it was so soon demoted to a goods station, Nine Elms has survived, at least in part. It established the U-shaped terminal plan with buildings and concourse behind the buffer stops and platforms extending from them to flank the tracks. In all the early termini of this type, the platforms and the lines serving them were arranged far apart and the space between them filled with sidings used for the storage of passenger rolling stock. This was a logical arrangement at the time, but as traffic grew additional platforms took the place of the stock sidings. This involved the removal of empty trains from arrival platforms to stock sidings situated at a considerable distance down the line, an uneconomical method of working.

Contemporary writers were critical of Hardwick's work at Euston, pointing to the contrast between the magnificent portico and the meagre, strictly functional train sheds to which it led. They therefore dismissed it as so much senseless extravagance. These critics missed the point. The portico was not merely the façade to a building, it was the entrance to a railway, a triumphal arch celebrating the completion of what was then man's greatest physical achievement. Nor was the humble train shed behind it so insignificant in what it portended, for the roof sheltering the platforms was supported on triangular trusses consisting of wrought-iron compression members of rolled T-section tied with rods of the same material. So far as is known, these were the earliest trusses or fabricated beams to be made wholly of wrought iron. In the following decades, the knowledge gained in the Britannia Bridge experiments was applied with increasing confidence to the construction at major stations of great vaults of iron ribs and glass arching over both platforms and tracks.

Because of the eclecticism of Victorian architects, opinion on the merits of their work is often sharply divided and judgement is apt to be swayed by nostalgia. On the merit of these great station roofs, however, there can be no difference of opinion, if only because functional considerations dominate their design, stylistic eclecticism being confined to details such as decorative cast-iron supporting columns. An architectural adventure of the greatest daring and skill, the soaring span of the Victorian train shed speaks for the nineteenth century and for no other. It is ironical that while Pugin and his disciples, inspired by the medieval cathedrals, were devoting themselves to recreating the pure Gothic style, the engineers were building these lofty pagan temples consecrated to the god of steam. There can be no question on which was the truer expression of an age dedicated to material progress.

For the first of these great stations, Newcastle Central, completed in 1855, the local architect John Dobson was responsible. If it had been built in accordance with his original design, this would have been the finest station building in the world, but unfortunately the need to accommodate the headquarters staff of the railway company entailed modifications which marred it somewhat. A recent writer suggests that Robert Stephenson must have assisted Dobson with the arched all-over roof, the first of its kind. But Dobson, as well as being an architect of rare talent, was also a capable engineer who, in addition to the roof, designed the special rolls used to produce the curving wrought-iron roof ribs. If he did collaborate at all it would have been with Stephenson's resident engineer, T.E. Harrison, who, as Stephenson acknowledged, was responsible for most of the work on the Newcastle & Berwick Railway.

Thanks to Dobson's versatility, Newcastle Central achieves a unity between functional iron roof and stylized masonry buildings that was seldom achieved elsewhere. If such a marriage was satisfactory it was usually where the engineer was dominant, as at Paddington or in the case of Lewis Cubitt's King's Cross. Of this last splendid design Cubitt himself remarked, with a good sense extremely rare for the period, that it achieved its effect through 'the largeness of some of its features, its fitness for its purpose, and its characteristic expression of that purpose'. Not for nothing did Cubitt belong to a family of engineers.

Generalizations are dangerous, but with rare exceptions one can say that the marriage of functional train shed and grandiloquent buildings became less happy as the nineteenth century wore on. Opinions may differ on the merits of Sir Gilbert Scott's huge Gothic building at St

Pancras, built to celebrate the Midland Railway's belated but triumphant arrival in London, but at least all must agree that it bears no architectural relationship whatsoever to engineer. W.H. Barlow's mighty parabolic iron vault which lies behind it. It could scarcely be otherwise when Sir Gilbert remarked, with a self-satisfaction almost incredible, that his building was 'possibly too good for its purpose'. An architect obsessed with Gothic detail was evidently totally blind to the merits of the engineer's splendid roof with its span of 240 feet, then the greatest in the world.

The 'new' stations at Bristol and York, completed within a year of each other in 1877–8, were the last of the great train sheds to be built in the grand manner. At Bristol Temple Meads it is possible to study, like successive layers in some archaeological dig, the whole range of Victorian station architecture. First comes Brunel's original terminus of 1840 with its stone buildings in Tudor style and its train shed with wooden false hammer-beam roof of seventy-two-feet span. With the descecration of Euston, this is now the finest monument of its kind and date left to us, but it is at present threatened with demolition. Upon the opposite side of the station approach stand the Jacobean-style offices of the old Bristol & Exeter terminus of 1852. Then comes Digby Wyatt's new through station of 1878 with its all-over roof of iron and glass, arching above the sharply curving platforms, and finally the extra platforms with their mean buildings and shelters added in the 1930s.

The history of York is very similar. The original station was a terminus within the city walls, where every passenger from the south bound for Darlington and points north must change. The new through station designed by Thomas Prosser and his successors replaced this terminus, though the original station survived until recently as offices. The largest station in the world at the time it was opened, the buildings of the new York station are undistinguished and insignificant, focusing all the attention on the three magnificent spans of the train shed. Like Newcastle and Bristol, York is set upon a curve but the roof is more extensive. To stand upon the central footbridge and look along the curving arcades of iron Corinthian columns from which the soaring roof ribs spring is to be in no doubt that this is one of the great buildings of the Victorian age.

In the design of the small country station, which usually included living-quarters for the station master, the railway architect's role was different. The aim here was not so much to impress as so appease; to convince a conservative country gentlefolk, by conforming to their own standards of taste, that the railway was respectable and knew its place. Local materials were used in the construction of station buildings which resembled a *cottage orné* or a lodge at the gates of some great estate. If the station served such an estate, its architect would be charged to produce something rather special such as a lofty *porte cochère*. Despite the wholesale demolition of wayside stations in recent years, some of these charming buildings still survive.

As the Victorian age wore on and people became accustomed to railways, there was less need for such costly appeasement and standardized station buildings began to appear, each company adopting its own highly characteristic form. These are no less interesting and it is unfortunate that, in our natural concern to preserve examples of the earlier, more individualistic station buildings, we may suddenly find that the later type, so typical of the pre-grouping companies, has all disappeared.

A pioneer of standardized buildings was David Mocatta, whose association with J.U. Rastrick on the London and Brighton line was one of the more successful of such collaborations between architect and engineer. Just how happy it was we can judge from the splendid viaduct over the valley of the Sussex Ouse north of Haywards Heath where Mocatta's balustrades and eight small flanking

Italianate pavilions add precisely the right civilizing touch to Rastrick's starkly functional engineering without being either fussy or pretentious. Mocatta built a fine station in the Italianate style at Brighton, but for the smaller stations on the line he evolved a basically identical form, achieving different styles by an inventive variety of detailing. In this variety, Mocatta's designs acknowledge the eclectic taste of 1841 when they were built, but in their standardized form they looked forward to a future when stylistic details would be standardized also, being increasingly governed by the patterns of cast-iron columns or brackets held in the railway companies' foundries at Derby, Crewe or Swindon.

Today, a man born in 1900 could argue convincingly that he has lived to see greater social changes than any previous generation. Yet it is doubtful whether a man born in 1800 would agree with him. He would say that the railways were a more potent instrument of social change than any subsequent invention and that the motor-car and the aeroplane are merely continuing a revolution that the steam locomotive began. For he was born into an ordered world where, despite the new roads and canals, life went on much as it had done for un-numbered generations. In the Midlands and the North, as a result of steam power and canals, the Industrial Revolution was already under way, but this new black world had not yet exploded over the green to any great extent. The Black Country was already blackening, but over most of the rest of the country the old static and parochial life went on as usual and people remained unaware of the new fires in the belly of the country. The new roads meant that mails and news circulated more freely, but coach travel was so expensive that only a very small minority could take advantage of it. For the vast majority, life revolved about the provincial market town. It was the centre of their world, a small and self-sufficient world in which local tradesmen and craftsmen catered for every need.

It was this world that the railway invaded with devastating impact, making thousands of Englishmen aware for the first time of the new civilization of coal and iron, of smoke and steam and furnace flame, of which they had remained blissfully unaware until it had suddenly put forth its iron tentacles. It is through the art of the period and not through its literature that we are made most vividly aware of the magnitude of this change. For as though men had a prevision of what it portended and of what they were about to lose, the Industrial Revolution coincided with the Romantic Movement that made Englishmen consciously appreciative, in a way they had never been before, of the unsullied beauties of the natural world. So they perpetuated in paint the landscape of that lost England on the very eve of its dissolution.

They also recorded for us the first agents of that disolution, for the railways with their many-arched viaducts and cavernous cuttings and tunnels appealed strongly to the Romantic imagination. It is these pictures of early railways that convey most eloquently the force of their impact, for they reveal the same eye for landscape and for man's immemorial way of life as a part of that landscape. Those women gleaning, that reaper with his sickle or the ploughman guiding the stilts of his wheel-less wooden plough might be figures from some medieval book of hours. But now they are portrayed working under the shadow of the towering piers of a great viaduct and they look up in wonderment at the train that thunders triumphantly overhead beneath a banner of smoke and steam. A group of Australian aborigines standing in front of a jet plane would not make a more striking contrast or one so poignant.

A man born in 1800 would have grown up in a pre-railway world that was nearer to medieval England than it is to our own. He would have seen the railways invade it and suck its life away into the growing manufacturing towns. By the time he reached his three-score years and ten his world had utterly changed. Except in a few remote backwaters untouched by railways, the stable, self-sufficient life of countryside and market town was dead. The railway had become an umbilical cord supplying all the necessities of life from large steam factories in the swollen industrial towns. Local tradesmen had either become mere shopkeepers or, tempted by the cheap fares on 'parliamentary' trains, had been sucked away to swell the industrial population. Railways made possible centralized manufacture upon an unprecedented scale; they also brought a new mobility. At first this was enjoyed mainly by the upper and middle classes, but soon cheap excursions brought railway travel within the reach of all. People who had never seen the sea thronged the piers of booming Victorian pleasure resorts or visited a London that had hitherto seemed as remote and unattainable as the antipodes.

The bustle at the inn as a crack coach changed horses or the familiar cry of the post horn that our man of 1800 remembered in his youth had become no more than a nostalgic memory, for in twenty years a great coaching industry had vanished utterly away to leave the trunk roads of MacAdam and Telford strangely silent. Instead of stables and coach-building shops there were vast railway workshops, each surrounded by a brash new town, Swindon, Crewe, Eastleigh, Doncaster, where blackened men wrought in iron instead of wood. Above all, the railways made Englishmen more time-conscious. Whereas before 'local time' had been good enough, efficient railway operation spelled the introduction of Greenwich Mean Time throughout the country. The England of 1870 must have seemed to a man as old as the century as fast-moving, as feverish and as fretful as our own. It was a new, strange world that the engineers had created for good or ill; the rest must ride the whirlwind.

TWO

Men of Steam

'I sell, Sire,' Matthew Boulton once remarked to George III, 'what all the world desires – power.' His words typify the pride and self-confidence that launched the Industrial Revolution. The world desired power and the steam-engine builders of Britain would supply it. Eighteenth-century pioneers like James Watt and Matthew Boulton believed that material progress, with the steam engine as its driving force, would lead directly to an earthly Utopia where human drudgery and poverty had been abolished and wealth undreamed of created by the unwearying power of the slave they had so successfully harnessed. The overwhelming conquest of the steam engine in the nineteenth century certainly proved the truth of Boulton's words, but though it created wealth it failed to eliminate poverty. It certainly removed much arduous physical toil, but only at the price of a different and worse kind of drudgery. In the new factories men's working lives were disciplined by the novel and unnatural rhythms dictated by tireless steam-driven machines.

James Watt did not, as is commonly supposed, 'invent' the steam engine, but greatly improved upon the engine designed and built in 1712 by his great predecessor, Thomas Newcomen. Newcomen's invention was, properly speaking, an atmospheric engine, using steam merely as a convenient method of creating the vacuum from which it derived its power. Steam at little above atmospheric pressure was admitted to the lower part of the cylinder below the piston where it was condensed by the admission of a jet of cold water. The cylinder was open-topped, and into the vacuum that was thus created the piston was driven down by the weight of the earth's atmosphere. This single-acting cycle of operation was extremely slow, making the engine unsuitable for rotative motion, and it was chiefly at a mine pump that it proved valuable. The piston and pump rods were connected by chains to opposite ends of a wooden overhead rocking beam.

The fact that the cylinder of the Newcomen engine was alternately heated and cooled during each working cycle made it very uneconomical. Nevertheless, it was particularly welcomed by the coal industry where its ability to keep mines clear of water and so enable deep-level seams of coal to be worked far outweighed its extravagant appetite for fuel. For this reason some Newcomen-type entines could be seen working at British collieries throughout the nineteenth century.

James Watt's most significant achievement was his discovery that the law of latent heat was the reason for the inefficiency of the Newcomen engine and that its cylinder should be kept constantly hot. To achieve this he invented his condenser, a separate vessel into which the steam from the cylinder was exhausted and there condensed. He also enclosed the top of the cylinder and substituted steam at low pressure in place of the cold air to act upon the piston. Thus although Watt's first successful engine was similar in appearance to its predecessor and was still only a single-acting non-rotative mine pump relying for its power on the creation of a vacuum, it was a true steam engine of much higher thermal efficiency. It also opened the way to further development and refinement by its inventor. The crude chain connections to the overhead beam were replaced by a positive linkage in the form of Watt's famous parallel motion. This, in turn, made the double-acting engine possible and with it rotative motion.

For the first time the Watt rotative engine offered manufacturers a reliable source of power other than the water wheel for driving machinery of all kinds, and from the date of its introduction in 1783 the factory system in Britain may be said to have begun. Like its predecessor, Watt's engine was slow-moving and ponderous and was not a self-contained power unit, being literally built in to its engine house so that the house formed part of the structure of the engine. But it was also extremely reliable and long-lived, so that, with subsequent modifications, in mines, ironworks and mills, pumping stations, factories and breweries, many Watt engines continued to work until the end of the nineteenth century and beyond.

Until 1800, the Watt engine was protected by a master patent which Boulton & Watt defended so energetically that further development was almost completely frustrated. During these years of enforced design stagnation, many engineers realized that the way ahead lay in harnessing the expansive power of steam at high pressure. Not only did 'strong steam', as they called it, promise higher efficiency, but a lighter, more compact power unit, pregnant with possible applications in the limitless field of transportation. But James Watt stubbornly opposed this school of thought, not because he questioned the power of high-pressure steam, but because he doubted the competence of existing technology to harness it with safety. Before condemning Watt's attitude as reactionary it is as well to remember that the typical boiler of the time was little better than a glorified brewer's copper.

It was in Cornwall that Boulton & Watt had the greatest fight to uphold their patent monopoly in face of the apostles of high-pressure steam. Despite successive booms and slumps in its mining industry, Cornwall at this time was the richest source of tin and copper in the world, a position it held until the 1870s when its mining industry collapsed in the face of competition from foreign ores.

In order to tap this mineral wealth, Cornish mines were driven ever deeper and there was a demand for increasingly powerful pumping engines to keep such deep levels clear of water. Moreover, the fact that all coal had to be imported at greatly increased cost supplied a unique incentive to improve the efficiency and economy of the steam engine.

There are said to have been seventy-five Newcomen engines at work in the Cornish mines in 1777 when the first Watt engine was introduced and its superiority was so quickly recognized that by 1783 there was only one Newcomen engine left there. But the same considerations that had made Cornish engineers Boulton & Watt's first and best customers in 1777 had transformed them into bitter rivals by the end of the century. It would be true to say that the high-pressure steam engine, the *primum mobile* of the nineteenth century, was born in Cornwall, the child of this rivalry. It took three forms: the simple expansion condensing engine, the compound expansion condensing engine and the simple non-condensing engine. Each type originated in Cornwall and their evolution will now be described separately.

The simple expansion condensing engine was evolved directly from the Watt engine which it outwardly resembled, the only difference being the higher working pressure. It could be either single- or double-acting. First of the type was a small Watt engine at Wheal Prosper mine in Gwithian which the Cornish engineer, Richard Trevithick, adapted in 1812 to work at a pressure of forty pounds per square inch instead of the 5–10 pounds used by James Watt. This was the origin of what became known to engineers throughout the world as the Cornish engine. Its most celebrated makers, Harveys of Hayle and the Perran Foundry, were Cornish, but the type was produced by many makers outside the Duchy and was renowned for its economy and reliability. The true Cornish engine was a non-rotative pump and its use became almost universal in the nineteenth century. Cornish engine-builders were happily cushioned against the collapse of the Cornish mining industry after 1870 because this coincided with the provision of water supply and sewerage systems in the larger towns which created an alternative demand for pumping engines of great power.

Throughout the century the Cornish engine underwent a continuous process of detailed refinement; the old wooden beams of the Newcomen and early Watt eras were replaced first by cast-iron and then, after 1870, by fabricated wrought-iron beams. Yet to the last it retained the same configuration as Newcomen's engine of 1712 and its direct lineal descent from this first steam engine in the world was obvious.

The Cornish engine attained immense size. That supplied to Battersea pumping station in 1858 had a cylinder 9 feet 4 inches in diameter, and engines with 90–100-inch cylinders were comparatively common. The beam (or 'bob' as it was termed in Cornwall) of such an engine might weigh anything up to fifty tons or more. In mine-pumping engines, the practice inherited from the earliest days was to support the trunnion bearings of these huge beams upon one extra thick wall of the engine house which was known in Cornwall as the bob wall. This meant that the pump end of the beam projected outside the house and over the mine shaft into which the pump rods descended. Such an arrangement was unacceptable in a waterworks pumping station where the engine had to be totally enclosed within the building. Consequently the beam had to be supported on massive trilithons of cast iron.

Whereas the strictly functional usually prevailed in the mining industry, the Victorian waterworks pumping station became the subject of increasingly elaborate architectural treatment which included not only the building but the engine itself. Usually the simple trilithon became a group of Doric columns supporting an entablature, but occasionally, as at Cross Ness, the Southern Outfall pumping station of the Metropolitan Sewage Works, there was a display of cast-iron Gothic of fantastic elaboration. Again, Whiteacre pumping station in Staffordshire presented, until its recent demolition, an extraordinary mixture of Egyptian-style columns and Gothic detail, the service platforms surrounding the cylinder heads being supported by gilded eagles. In such lofty, extravagantly decorated halls, incensed with the smell of steam and hot cylinder oil, seen by few people other than their attendants, the great engines laboured night and day for seventy years or more. When the Metropolitan Water Board was formed in 1884, it inherited 150 engines from its eight constituent companies and of this number twenty-five Cornish engines were still in use in 1927. One of the longest lived of these is now preserved at Kew. A ninety-inch engine built by the Copperhouse Foundry, Cornwall, it started work at Kew on 30 May 1846 and worked until 31 July 1943.

Beam engines of similar type were also widely used as blowing engines to provide blast for the furnaces of the booming Victorian iron industry. For obvious reasons, survivors are rare. Although they are no longer used, a pair of these beam blowing engines, built by Murdock and Aitken of Glasgow in 1851, and named *David* and *Sampson*, survive in the Lilleshall Ironworks in Shropshire.

Simple expansion rotative beam engines, known in Cornwall as 'whim engines', were widely used for mine and colliery winding and were generally smaller than the mine-pumping engines as the power requirement was less. In this application they were superseded in the colliery districts later in the century by winding engines of more specialzed design, but in other industrial applications where the power demand was high the rotative beam engine persisted and some were eventually modernized as we shall see presently.

None of this development would have been possible without boilers capable of generating – and withstanding – steam at high pressure. The first such boiler was the

Cornish type which Richard Trevithick is said to have originated in 1812. Built from wrought-iron plates riveted together, it was of cylindrical form with a single large central flue running through it and containing the furnace grate at one end. It was widely used as a steam supplier for stationery engines until 1844 when William Fairbairn and John Hetherington of Manchester introduced the Lancashire boiler. This was similar to the Cornish type except that it had two parallel flues, each with a furnace, instead of one. Fairbairn's object in introducing this boiler was that be the use of two furnaces, fired alternately, less smoke would be produced, but the additional heating surface provided by its two flues generated so much more steam than did the Cornish type that the Lancashire was quickly and widely adopted and became the most popular boiler of all time. Successive improvements in construction enabled it to meet the slow but steady increase in pressure demanded throughout the Victorian age and even today it is still widely used to generate process steam.

The alternative to using high-pressure steam expansively in a single cylinder is to allow it to expand partially in one cylinder and then to enable it to pass to a second, larger cylinder where the process of expansion is continued before the steam is finally exhausted to a condenser. This is the principle known as compound expansion and it was first introduced in 1791 by the Cornish engineer, Jonathan Hornblower, in an attempt to evade the Boulton & Watt patent monopoly. Although Hornblower's engine proved little superior to the Watt type, his attempt to introduce it led to a protracted legal battle. In 1811 a London engineer named Arthur Woolf introduced a similar compound design to Cornwall, but although he built a number of engines, their improved efficiency was insufficient to justify the additional cost and complication of two cylinders. The reason for this comparative failure was that at the steam pressure of forty pounds or so which was then the rule, compound expansion was not justified. It was when pressure had climbed to 100 pounds or more that the compound engine came into its own in Britain through the agency of William McNaught of Manchester.

By 1845 many factories and mills were finding that, even with higher steam pressures, their beam engines, many of which were of the old Boulton & Watt type, were no longer equal to the increasing demand for power that new machines and methods made on them. They were therefore facing the costly prospect of replacing them. McNaught supplied an economic alternative. He converted the existing engines to compound expansion by adding a high-pressure cylinder exhausting into the original low-pressure cylinder. The new cylinder was connected to the opposite end of the beam midway between its pivot and the connecting rod end so that its stroke was necessarily shorter than that of the original cylinder. This conversion proved completely successful and a large number of steam engines, particularly in the textile mills of Lancashire and Yorkshire, were altered in this way so that the term 'McNaughting' passed into the language of steam engineers.

It was from 1845 onwards that compound expansion began to establish itself and an increasing number of new engines were built on this principle. These were of three types: first, with cylinders side by side,* secondly with cylinders in line, sharing a common piston rod (tandem compound) and thirdly and more rarely with the low-pressure cylinder enclosing the high-pressure (annular compound).† The most notable of these early compound engines were of the last-mentioned type and had cylinders

* When built in horzontal form with central flywheel between the cranks, this type was known as a 'cross compound'.
† The low-pressure piston was of annular form, encircling the smaller cylinder, high-pressure steam being admitted below its piston and then passing to the tops of both cylinders on the downstroke.

twelve feet in diameter, the largest ever made. Three of these huge engines were built by Harveys of Hayle between 1843 and 1849 and supplied to Holland for draining Haarlem Mere. As this mere was seventy square miles in area and was estimated to contain 1,000 million tons of water, the task must have seemed a labour of Hercules. However, when all three engines were set to work they lifted 2.8 million tons of water a day and, despite the fact that storm damage to the sea defences in January 1852 undid the previous three years' pumping, the mere was drained by the end of 1855. After this the three engines were occasionally used to maintain the correct level in the drainage canals. One of them, the Cruquis engine, last worked in June 1933 and is now preserved on site as an outstanding example of British steam engineering in the early Victorian period.

In 1800, when Richard Trevithick designed and built for the Wheal Hope mine in Cornwall a small steam winding engine which became known locally as a 'puffer whim' because it used high-pressure steam and exhausted it to atmosphere instead of to a condenser. To James Watt and the rest of the steam engineers of the day this seemed sheer heresy, for all engines at that time used condensers and relied upon the creation of a vacuum below the piston for a considerable proportion of their power. Even Trevithick's fellow-Cornish engineers, though they advocated the use of steam at higher pressure, never dreamed of throwing away the condenser; only Trevithick had the courage to do this. He realized that the advantage of the condenser could, at best, only amount to one atmosphere, or approximately fourteen pounds per square inch, and that, in a condensing steam engine, this advantage was partly discounted by the power absorbed in working the pumps which maintained the vacuum in the condenser. By using steam at the pressure of several atmospheres, the loss of the condenser would be more than offset by the advantages of simplicity, lightness and portability.

So, on Trevithick's new 'whim engine' there was no condenser with its attendant pumps; nor was there any ponderous beam. The cylinder was let into the top of the boiler, its piston rod driving the crankshaft and flywheel directly via a cross-head and connecting rods. The engine was delivered to the mine complete in an ordinary farm cart at a cost of 10s. 6d. For the first time man had produced a compact and portable source of power. Its possibilities were immense, as Trevithick himself was the first to realize, but unfortunately his ideas fell on stony ground. The road vehicles and, more particularly, the locomotives which he built on the lines of his 'puffer whims', though they pointed the way for others, met with no commercial success. Colliery-owners in South Wales and Tyneside did not consider that the capital cost of strengthening their fragile wooden or cast-iron tramways to stand the weight of Trevithick's locomotive would be justified. So they carried on with horse haulage until the great increase in the price of horse fodder brought about by the Napoleonic war caused them to think again. Then George Stephenson and other Tyneside engineers began building locomotives on the lines laid down by Trevithick.

Robert Stephenson was the man primarily responsible for developing these first crude coal-hauliers into the machine that made the railway age possible. He achieved this in an astonishingly short space of time between 1828 and 1833 at the works of Robert Stephenson & Co. of Newcastle, the first locomotive-builders in the world. In his famous *Rocket* and the similar machines which followed it, the cylinders were arranged beside, instead of within, the boiler, driving the wheels directly and eliminating the overhead crossheads. But the outstanding feature of the *Rocket* was its boiler, which was the prototype of all locomotive boilers built subsequently. Earlier types had resembled Trevithick's Cornish boiler in having a single central flue or, at best, a U-shaped return flue. For this, Stephenson, at the suggestion of Henry Booth, substituted many small fire tubes connected to a water-jacketed firebox. By thus greatly increasing the heating surface, he ensured that sufficient steam could be generated for continuous running. Trevithick had earlier shown that the draught on the fire could be quickened by turning the exhaust steam into the chimney, but this principle of the steam blast only became fully effective when applied to the multi-tubular boiler.

In 1830, Stephenson produced his *Planet* locomotive, a four-wheeled design (2–2–0) in which the two cylinders were mounted directly beneath the smoke box and drove a cranked rear axle. Thus cylinders, connecting rods and valve gear were all inside the frames, producing a very neat and compact layout. This *Planet* design was soon enlarged in two alternative six-wheeled designs, the *Patentee*, with single central driving axle (2–2–2), intended for passenger service, and the *Atlas* in which all six wheels were coupled (0–6–0). Thus, before Victoria came to the throne, there had been evolved the two classic types of locomotive which, with but few exceptions, handled all the traffic over Britain's expanding railway system until the 1860s. Indeed, the six-coupled engine remained the standard British goods locomotive throughout the nineteenth century.

The Achilles heel of the early inside cylinder locomotive was the cranked axle. Until technique improved in the second half of the century it was difficult to forge a satisfactory cranked axle that was free from flaws. Hence the breakage of such axles was a frequent cause of locomotive failure. So that such a failure would not cause a major disaster, Robert Stephenson adopted a double frame design which provided bearings on each side of the driving wheels to prevent them from collapsing if the axle broke. Daniel Gooch, who worked in his youth for Robert Stephenson & Co., brought this double frame design to the Great Western Railway where it was perpetuated until the early years of the twentieth century. Other Victorian locomotive engineers evaded this axle problem by adopting outside cylinders driving crankpins on the wheels. This was similar arrangement to the famous *Rocket* except that the cylinders were placed on either side of the smoke box instead of at the rear.

In 1842, two employees of Robert Stephenson & Co., William Williams and William Howe, designed an inspired simplification of the complex 'gab' valve gear hitherto used on the *Patentee*. This became known as the Stephenson link motion, and it was an invention as important as Watt's earlier parallel motion. Like the latter, it was a simple and beautiful geometric arrangement of rods and links and it enabled the driver not only to reverse the locomotive by the movement of a single lever but, by adjusting the same lever, to vary the moment when the steam supply to the cylinder was cut off. When starting from rest or working heavily, steam admission could be prolonged, but when running easily at speed the steam could be cut off earlier and thus used expansively. This effected a great economy in steam consumption and therefore in fuel. So great were the advantages of the Stephenson link motion that it was speedily adopted on all locomotives and at the end of the century it was still the most popular type of valve gear.

By 1845 the steam locomotive had become a highly refined machine, fast, reliable, powerful and exhibiting most of the features destined to persist to the end of the steam era on railways. The main reason why early locomotives appear so archaic to us is the absence of any form of protection for the engine crew. The omission was deliberate. Most engineers agreed with Brunel's expressed view that to pamper the engineman by giving him the shelter of a cab would be to invite accidents because his attention would wander. In other words, there was nothing like unlimited fresh air, with a liberal addition of steam, smoke and smuts, for keeping men alert. After all, the coachman had thrived on almost total exposure for long enough so why should his successsor be given special treatment?

One salutes those enginemen of long ago who, clad in their white corduroys, may be seen in early photographs posing proudly on their open footplates. Imagine what it was like to drive such a locomotive, even at the modest speeds then achieved, through a heavy storm of rain, hail or snow, probably at night when only the ability of the engineman to pick up the dim warning lights of the signals could avert swift and sudden disaster. It was soon realized that in the responsibility he bore and the conditions he had to face there was no fair comparison between the engine-driver and his predecessor, the coachman. So, in the 1840s, the spectacle plate made its appearance. This was a more or less inadequate metal windscreen provided with two small circular windows. From the spectacle plate the earliest vestigial cabs developed, first by the addition of side sheets and then by the provision of a narrow roof which seldom sheltered more than a quarter of the footplate. Strangely enough, the evolution of the cab was resisted by the rugged enginemen themselves who argued that it would obstruct their view ahead. It was not until 1886 that T.W. Worsdell, the locomotive engineer of the North Eastern Railway, introduced a roomy cab with side windows and a roof extending right over the footplate. Other lines, such as the Great Eastern, soon followed suit, but some, like the Great Western and Great Northern, retained inadequate cabs until well into the twentieth century. Indeed, those now in middle age may recall the tarpaulins, extending from the cab roof to the front of the tender, which Great Western enginemen used to unfurl for additional protection in wet weather.

Despite the fact that its designers paid scant regard to the comfort and convenience of its crew, the steam locomotive rapidly became, and remained throughout the century, the most outstanding example of Victorian mechanical engineering by whatever technical or aesthetic standard it be judged. It passed rapidly through three phases of evolution. First, there was what might be called the primitive blacksmith's style of the first coal-hauliers in which the boiler was surmounted by a crudely functional clutter of moving rods and levers. This was followed by the neat simplicity of locomotives of the family of Stephenson's *Patentee* in which the classical style was evident in elegantly fluted steam dome casings or safety valve columns. In this phase, the locomotive assumed, a little incongruously perhaps, some of the architectural features with which the large stationary engines of the period were embellished. In the final phase, which lasted until the end of the century and beyond, while the neatness and simplicity of design were retained, there was a return to a strictly functional form in which ornament was confined to polished metal beading or chimney cap or to the openwork of 'paddle box' splashers over great driving wheels.

Whereas European designers festooned their locomotives with extraneous pipes, sand domes, brake cylinders and other devices, a practice which American engineers adopted towards the end of the nineteenth century, the great Victorian locomotive engineers of Britain continued rigorously to eschew such unsightly ironmongery. It is arguable that such studious concealment of functional parts made their machines more difficult and costly to maintain. That by sheer design artistry in line, curve and proportion they endowed the British steam locomotive with a beauty explicit of power and grandeur such as no other machine has ever equalled cannot be denied.

Later in this book we shall see how, for a variety of closely associated reasons, the skill, artistry and pioneer spirit of the early Victorian engineers was lost in the second half of the century. To this trend, British locomotive engineers were a shining exception. They may have erred on the side of conservatism, but the majesty of their handiwork continued to reflect the pride and the craftsmanship of a great tradition. The reason for this exception is clear. Throughout the century the locomotive works of the great railway companies at Crewe, Derby, Swindon, Eastleigh, Darlington or St Rollox were little kingdoms over which successive Locomotive Superintendents (or Chief Mechanical Engineers, to use the later title) ruled with an authority that was almost absolute. Provided the railway company's demand for motive power was adequately met, its directors never interfered and even when it was not they sometimes hesitated to do so. A notable example of this was the situation which developed in the 1880s on Britain's largest railway, the London & North Western, which proudly styled itself the 'Premier Line'.* Francis Webb, at that time the uncrowned king of Crewe, decided to introduce compound expansion and designed a whole series of locomotives in which the two cylinders outside the frames exhausted into a single large low-pressure cylinder between them. The experiment was not a success. Performance was poor, coal consumption high and the locomotives suffered from a number of unfortunate defects, the most embarrassing being a tendency for the four uncoupled driving wheels to revolve in opposite directions on starting. Everyone on the railway from footplatemen to Chairman and General Manager was only too well aware of the shortcomings of these Webb compounds, yet their author remained blind to their defects and such was his autocratic position that the line struggled on with his compounds until 1903 when Webb finally retired. But such lapses were rare.

In few other spheres of activity did engineers retain such undisputed sway so long and in this way an ethos from the first pioneer days was passed from one engineer to his successor. Matthew Kirtley of the Midland Railway worked

*The Liverpool & Manchester and the Grand Junction were among its constituent companies.

as a boy on the Stockton & Darlington, as a fireman on the Liverpool & Manchester, and, as a driver on the London & Birmingham, is said to have driven the first main-line train to enter the capital. He was responsible for founding the Midland locomotive works at Derby where he reigned until his death in 1873 when he was worthily succeeded by Samuel Johnson. Edward Fletcher, locomotive superintendent of the North Eastern Railway for nearly thirty years, was a pupil of Robert Stephenson who assisted in the first trials of the *Rocket* and later drove the first Stephenson locomotive *Invicta* on the Canterbury & Whitstable Railway. Such were the men who nursed and handed on a great tradition.

Trevithick's first high-pressure 'whim engine' from which the locomotive stemmed also gave birth to the smaller, lighter stationary high-pressure steam engine which, for all but the highest power requirements, began to supersede the beam engine from the 1840s onward. Whereas the big beam engine had to be built on the working sie, the smaller high-pressure engine was a self-contained unit and therefore possessed the great advantage that it could be assembled complete and tested by its maker, even if it had to be subsequently dismantled for delivery to the customer.

Boulton & Watt's great rival, Matthew Murray of the Round Foundry, Leeds, was the pioneer of the commercially built self-contained steam engine in the first decade of the nineteenth century. He produced a number of small (by the standards of the time) engines of this type. Some were orthodox beam engines but with the beam supported on two integral 'A'-frames of cast iron. Others were inverted beam engines and a third type had a vertical cylinder driving directly to a crankshaft* and fly-wheel above it.

*Murray's first vertical engines did not employ a crank but used instead his ingenious arrangement of gears known as the 'epicycloidal motion'. An early Murray engine of this type may now be seen in the Museum of Science & Industry, Birmingham.

The steam locomotive developed so rapidly during the first half of the nineteenth century that it influenced the design of the stationary engine from which it sprang. Locomotive type slide valves and Stephenson link motion were widely adopted and the simple horizontal configuration of the inside cylinder locomotive was mounted on a substantial bed to become an increasingly popular power unit of single, duplex or compound type. Large horizontal engines, usually single crank tandem or cross compound, succeeded the beam engine as the most popular unit in the textile mills of the north, usually in conjunction with rope drive to the machinery on the various floors of the mill. Many of these engines continued to work until recently and a few still survive. The vertical engine pioneered by Matthew Murray was developed by William Fairbairn, Phineas Crowther and others. It was used for colliery winding, particularly in Durham, and was also popular for driving machine shops where its crankshaft could be coupled or geared directly to the line shafting so that the engine occupied little space in the workshops.

What we now regard as the orthodox type of vertical engine with cylinder above the crankshaft appeared later in the nineteenth century when it was originally called an 'inverted vertical' to distinguish it from the older type.

Cylinders and other vital parts for the earliest steam engines had all been produced by ironworks; by the Coalbrookdale Company, by John Wilkinson's ironworks and Bersham, Brymbo or Bradley, by the Carron Company in Scotland or by the Butterly Company in Derbyshire. At a time when transport was difficult it was logical that such heavy components should be made in the places where iron was produced. Better transport, however, made possible a new kind of engineering shop specializing in the manufacture of steam engines and other machinery. Boulton & Watt's Soho Foundry and Matthew Murray's Round Foundry at Leeds were the first examples of this new type of undertaking and were made possible by water transport. In the railway age they multiplied greatly. Not only did the railway companies build their own workshops, but numerous independent engineering shops supplying locomotives, steam engines and engineering products of all kinds sprang up, particularly in the areas around Glasgow, Newcastle, Leeds and Manchester. Mechanical engineering thus became an increasingly important industry.

Until 1847, the Institution of Civil Engineers was alone representative of the profession in all its manifestations. At the time it was founded, the term 'civil engineer' did not mean what it does today but was used to denote any engineer in civil, as distinct from military, employment. Nevertheless, it had been founded in 1818 by the builders of canals, new roads, bridges and harbours and this interest was still dominant in 1847. Mechanical engineers, most notably the new locomotive men, felt that their increasingly important craft was not adequately represented or catered for so they banded together and formed the Institution of Mechanical Engineers. Its first headquarters were in Birmingham, but in 1877 these were moved to London where they were eventually established within a stone's throw of the senior Institution. George Stephenson was the first President of this new Institution whose members included representatives of all the principal engineering and locomotive works throughout the country from Robert Neilson of the Hyde Park Works in Glasgow to Nicholas Harvey of the Hayle Foundry in Cornwall.

This development was symptomatic of the rapid growth of the engineering industry during the Victorian age. As it grew, so engineering knowledge ramified until it became no longer possible for a single engineer to bestride the whole profession in the way Brunel, Robert Stephenson or Locke had done. Their successors were forced to specialize ever more narrowly. As a result, many other professional Institutions representing specialized branches of engineering would be founded before the century was out. But for the present the railway-builders and the men of steam dominated the Victorian engineering scene.

THREE

Smoke Over the Sea

The replacement of sails by steam power at sea during
the Victorian era had results as momentous as the
introduction of railways on land. Just as the speed of
the railway train shrank the continents so the certainty of
the steamship drew them closer together. At the beginning
of Victoria's reign, the perils and uncertainties of a long
ocean voyage under sail were so great that few landsmen
lightly undertook such a journey and, having reached
their destination in safety, fewer still ever returned. By
the end of the century, a return trip from Southampton
or Liverpool to New York had become a comfortable and
commonplace experience for those who could afford it.
In this rapid revolution in ocean communication British
marine engineers and ship-builders played an outstanding
part. The industry they founded became the greatest in the
world, so much so that by the end of the nineteenth century
five out of every six ships to be found on the sea lanes of the
world were British-built.

The steamship has an ancestry older than that of the
steam locomotive though its rate of introduction was slower.
Primitive experimental steam vessels had appeared on the
sheltered waters of rivers or canals in England, France and
America before the eighteenth century was out. This was
because a ship's hull could, however inconveniently, accommodate the primitive and bulky low-pressure steam plants of
the time. The orthodox form of beam engine, though it was

used with success on such craft as the Mississippi steamboats, put the centre of gravity of the vessel far too high for safety in any seaway. Although a single stern paddle-wheel (as used on the Mississippi) was employed on some early steam craft, two side paddle-wheels mounted on the ends of a cranked paddle-shaft lying athwartships soon became the orthodox method of propulsion. This imposed on the marine steam-engine-builder a configuration of vertical or inclined cylinders with connecting rods coupled to the cranks of the paddle-shaft overhead. With the large, low-pressure engines then in use it became a task of great ingenuity to design a sufficiently power-ful engine within the limited vertical dimension between the paddle-shaft and the bottom of the hull and at the same time to keep the bulk of the weight as low down as possible.

One solution to this problem was the so-called 'side-lever' engine in which twin beams were disposed beside the vertical cylinder, the piston rod being coupled to them by crossheads and tandem connecting rods. Alternatively, two types of direct-acting marine engine were evolved. One of these was the trunk engine in which, as in the modern car engine, the connecting rod was directely attached to the piston, thus dispensing with the need for a piston rod and crosshead, and so reducing overall height considerably. To achieve this, the skirt of the piston was extended in the form of a hollow trunk, working through a gland in the cylinder cover, which gave the engine its name. The alternative type achieved the same result by mounting the cylinders in trunnions and allowing them to oscillate, covering and uncovering the steam ports as they did so, on the principle used to this day on many toy steam engines. With this arrangement the piston rod could be directly connected to the crank.

Although originally conceived for marine use, side lever, trunk and oscillating engines were all adapted for industrial purposes. Sometimes a more orthodox direct-acting arrange-ment was used to drive paddle-wheels by inclining two cylinders diagonally, both driving a single crank.

During the first half of the nineteenth century, many of the more celebrated makers of marine steam engines were based in the vicinity of London's river: Henry Maudslay and his partner Joseph Field of Lambeth whose side-lever engines were judged the best and most reliable marine power units in the world in the 1830s; the brothers Jacob and Joseph Samuda of Blackwall; John Penn of Greenwich; John Hall of Dartford who developed the trunk engine and, last but not least, Robert and David Napier of Glasgow and Millwall who, more than any other men, were responsible for the success of the steamship on the short sea routes.

The first steamship in the world to enter regular sea-going service was the *Rob Roy* of ninety tons burden. Built by Wlliam Denny of Dumbarton and launched in 1819, her hull measured eighty feet by sixteen feet and was powered by a thirty-two h.p. engine built by Robert Napier. She plied between Broomilaw and Belfast via Campbeltown. She was later transferred to the Dover-Calais service where she so impressed the French Government that they purchased her for their cross-channel mail service, renaming her, first the *Henri Quatre* and finally the *Duc d'Orléans*. The British Post Office soon followed suit with their own Dover mail steamers *Fury*, *King George* and *Eclipse*.

Meanwhile on the Clyde the *Rob Roy* was succeeded by the lager *Superb* (120 by seventy feet, 246 tons) built by Scotts of Greenock and launched in June, 1820. Again, this ship was engined by Napier. With a sister-ship, the *Robert Bruce*, the *Superb* began operating a twice-weekly service between Greenock and Liverpool.

By the time Victoria came to the throne, services of paddle-steamers, or 'steam packets' as they were called, were operating cross-channel services between England and the Continent or Ireland on a number of different routes. Dover and Milford Haven were two early examples of steam-packet ports, originally served by road coach, that received a great fillip when the railway reached them.

Dover has retained its supremacy as a channel port, but Victorian concern to minimize the perils of *mal de mer* at any cost would lead to changes in the Anglo-Irish sea routes. Railways were extended at heavy cost in order to shorten sea crossings. Stranraer, Heysham, Holyhead and Fishguard respectively in the Irish passenger service, while the similar railway-sponsored 'packet ports' of Larne, Greenore, Kingstown and Rosslare appeared on the other side of St George's Channel.

The growth and prosperity of the great Thames side ship-building industry had been due to the proximity of the oak forests of the Weald. So long as ships' hulls were built of oak, as all the early paddle-steamers were, Thames ship-builders continued to prosper, but the changeover from wood to wrought iron in ship construction threatened them with extinction because the sources of the raw material were all in the north. Hence the nineteenth century saw this great Thames industry virtually wither away and the corresponding rise to ascendancy of the Clyde and the Tyne. One firm, the Thames Ironworks, struggled on until 1912 when the building of the last of the many great ships to be launched into the Thames, the battle cruiser HMS *Thunderer*, drove them into bankruptcy.

One of the first men to see which way the wind was blowing was David Napier, but few were so far-sighted as he. 'London never will be a place for building steamers on account of everything connected with their production being higher there than in the North', he wrote when he retired, and his sons, prudently taking his advice, decided to concentrate the Napier interests in Glasgow in 1852.

Men began building ships' hulls of iron long before they understood the properties of the material. John Wilkinson, the ironmaster, was responsible for building the first iron boat in 1787. He successfully launched it into the Severn at Coalbrookdale to the disappointment of a large crowd who expected it to sink like a stone. Curiously enough, the first iron steamship that ever put to sea was built at Tipton in the Black Country, about as far away from the sea as one can get in England. She was constructed by the Horseley Coal and Iron Company, a polytechnical concern after the model of the Coalbrookdale Ironworks which had been founded about 1781, and she was named the *Aaron Manby* after her designer, a Shropshire man who was then the managing partner of the Horseley Company. The iron hull was 106 feet long by seventeen feet beam with a thirty-two h.p. engine driving side paddle-wheels which added seven feet to the beam. When completed, she was dismantled and despatched in sections by canal to London where she was reassembled at Surrey Canal Dock. She ran her first trial between Blackfriars Bridge and Battersea on 9 May 1822. Intended for river service on the Seine between Le Havre, Rouen and Paris, she left London in the middle of the month with a cargo of 116 tons of linseed and iron castings consigned to Paris, arriving safely at Rouen on 27 or 28 May.

Incidentally, Aaron Manby not only had an interest in the Seine steamboat service, but established a large and flourishing ironworks at Charenton by the junction of the rivers Seine and Marne near Paris. Staffed with skilled English employees, this works had a great influence upon French industrial development, but Manby was severely taken to task at home for enticing skilled men away from Britain. Concern about the 'brain drain' is no new thing.

Another early iron ship of Manby's design was built at Horseley to the order of Charles Wye Williams, manager of the City of Dublin Packet Company, who had a second iron paddle-steamer built at Cammell Laird's shipyard at Birkenhead, the first iron ship to be built by that company. Both these steamers were used on the river Shannon, where the hull of the second, the *Lady Lansdowne*, was recently discovered under water at Killaloe.

From this date forward the iron-hulled steamship very slowly gained ground. The shipwright's craft was an ancient one, steeped in tradition, and it was not easy for such an old dog to learn new tricks. Thus the first iron hulls were built on wooden-ship lines with a multiplicity of transverse frames (which had been necessary to hold the planking) and no advantage was taken of the inherent strength of the iron plating itself. This was not realized.* Thus the development of the iron hull owed more to the experience of boiler-makers than to ship-wrights.

The first and most convincing demonstration of the strength of an iron hull occurred in the spring of 1845 during the launching of the iron steamer *Prince of Wales* from Miller & Ravenhill's yard at Blackwall. Owing to the breakage of the launching gear, the bow of the ship became lodged on a wharf, and in the process of pushing it off with jacks, the hull was left unsupported over a length of 110 feet. In such circumstances a wooden ship would have broken her back, but the shipwrights were astonished to discover that the iron hull had sustained because the midship section had been left undecked to allow the engine and boilers to be taken on, thus denying the hull vital stiffening precisely where it was most needed.

This mishap and the lesson learned from it had far-reaching results. It encouraged Robert Stephenson to pursue his ideas for his Britannia Bridge. Moreover, William Fairburn, who had taken over the ship-building and engineering business of Ditchburn & Mare at Millwall, was an eye-witness of the occurrence and this greatly increased his confidence in the strength of wrought iron. It therefore led directly to his classic series of experiments at

---

*There is an interesting twentieth-century parallel here in the evolution of the motor-car body. In early wooden-framed bodies no advantage was taken of the potential strength of the metal panelling which replaced the earlier wooden panelling. In modern 'monocoque' construction the metal body itself forms the backbone of the car.

Millwall which not only gave birth to the Britannia Bridge but resulted in the much more intelligent use of iron in ship-building.

The first steamship to cross any ocean was the American *Savannah* in 1819, but she used steam power only as an auxiliary to sail and on a voyage of twenty-seven days eleven hours from Savannah to Liverpool she steamed for only eighty-five hours. Throughout the next fifteen years a number of similar voyages were made but they did nothing to dispel the widely-held belief that an ocean crossing under continuous steam power was impossible because no steamship could carry sufficient coal for such a voyage. It was not until 1838, the year after Victoria's accession, that the falsity of this notion was triumphantly demonstrated by I.K. Brunel.

Such was the versatile genius of this most remarkable man that in the midst of his hectic career as a railway engineer he was able to design three outstanding steamships, each of them the largest ever built at the time of launching and each a notable milestone in design. The story began in October 1835 at the time the railway from London to Bristol was being built, when Brunel suggested to the astonished directors, most of whom thought he was joking, that the railway might be 'extended' by building a steamship named the *Great Western* to ply regularly between Bristol and New York. But Brunel spoke in earnest. He had realized that while the cargo capacity of a ship's hull increases as the cube of its dimensions, its resistance to motion through the water (and therefore the power required to drive it) only increases as the square of those dimensions. Therefore, he argued, to design a steamship for any given length of voyage became simply a question of determining the right proportions. This simple proposition, which does not appear to have occurred to any engineer before, governed the design of all three of Brunel's great ships.

Although the Board of the Great Western Railway fought shy of Brunel's idea, certain Bristol merchants became convinced of its soundness and formed the Great Western Steamship Company. Construction of the ship began at Patterson's Yard in Bristol on July 1836. The hull of the *Great Western* was built of oak by traditional methods, but apart from its unusual size (overall length 236 feet; breadth over paddle-boxes 59.8 feet) immense strength was built into it, particularly in the longitudinal plane, the better to withstand the storms of the north Atlantic. For her engines, Brunel insisted upon the best that were then available – a pair of side-lever engines of 750 i.h.p. by Maudslays – and the ship was sailed round to the Thames to take them aboard and to complete her fitting-out.

As soon as it was realized that Bristol and Brunel meant business, the two rival ports of London and Liverpool laid down challenges for the north Atlantic route, and when it became clear that their new ships would not be ready in time, both ports resorted to charter. In the event, however, only the little *Sirius*, representing the British & American Steam Navigation Company of London, was available in time to outbid the *Great Western*. She was a little Irish sea steam-packet boat which was modified for the occasion by the removal of part of her accommmodation to make way for additional coal bunkers.

On 28 March 1838, the *Sirius* sailed from the Thames for New York, via Cork where she intended calling to replenish her bunkers. *The Great Western*, with Brunel on board, followed in her wake on 31 March, bound for Bristol, but, alas, fire broke out on board when she was off Leigh-on-Sea due to the boiler lagging having been carried too close to the hot base of the funnel. The fire was successfully extinguished but, in fighting the flames, Brunel was seriously injured and had to be put ashore at Canvey Island. Meanwhile the Master, Lieutenant James Hosken, R.N., fearing the worst, had beached his ship on the Chapman Sand where she had to wait until the tide floated her off. Because of this unlucky mishap, the *Great Western* did not reach Bristol until 2 April and could not leave there for New York until the eighth of the month, four days after the *Sirius* had sailed from Cork. It seemed that the Atlantic race had been lost before it had properly begun, yet it was to be a close run.

To hazard a small cross-channel steamer on the north Atlantic so early in the year was an heroic venture, but the little *Sirius* battled on and eventually docked at New York on 23 April after nineteen days at sea. As the first ship to cross the Atlantic under continuous steam she richly deserved the enthusiastic reception she received, but as she had to eke out her coal by burning part of her cargo and had only fifteen tons left when she docked, her feat scarcely proved the practicability of ocean steam navigation. In this regard there could be no doubt that Brunel was the moral victor in this epic contest.

For hardly had the *Sirius* docked before it became known that the *Great Western* was at anchor off Sandy Hook awaiting the pilot. Her voyage from Bristol had taken fifteen days five hours and, what was far more significant, she still had 200 tons of coal in her bunkers. No wonder she received a rapturous welcome from New Yorkers when she docked. At the sight of her 'huge size' and 'magnificent proportions' the crowds broke into spontaneous cheering.

> Myriads were collected, [wrote one of the *Great Western's* passengers] boats had gathered round us in countless confusion, flags were flying, guns were firing, and cheering rose from the shore, the boats and all around loudly and gloriously as though it would never have done. It was an exciting moment, a moment of triumph.

From this time forward the era of regular ocean steam services began although the first small paddle-steamers were ill-equipped to weather the worst storms and services were usually withdrawn during the winter months. But this precaution did not prevent the loss of the steamer *President* which foundered in mid-Atlantic with the loss of all hands in 1840. Notwithstanding this and several similar losses, the *Great Western* continued to ply regularly to New York, completing sixty-seven crossings in eight years. She was not only the safest but the fastest ship of her day, her best crossing occupying twelve days six hours.

The *Great Western* had scarcely completed her second Atlantic voyage before Brunel began to plan another and larger wooden paddle-steamer of not less than 2,000 tons burden. A visit to Bristol of the little iron-hulled steamer, *Rainbow*, however, encouraged Brunel to change his plans and adopt an iron hull of unprecedented size. Then, in 1840, Brunel changed his mind again by deciding to use screw propulsion instead of paddle-wheels for the new ship.

Because the Archimedean screw had been used as a water-lifting device by the Greeks before the birth of Christ, it may seem surprising that its use for ship propulsion did not take practical form until 1838, and then only after forty years of experiment. In this year, two practical forms of ship propeller were invented simultaneously by Sir Francis Pettit Smith and by a Swedish engineer resident in England named John Ericsson. The latter was a gifted engineer whose locomotive the *Novelty* had competed against the *Rocket* at the Rainhill trials. He left for America in order to exploit his propeller there, leaving the English market to Pettit Smith.

The first successful screw steamer in the world was the *Archimedes* of 237 tons, launched by the Rennie Brothers from their yard at Millwall in November 1838. The Great Western Steamship Company chartered her for six months to enable Brunel to carry out experiments and observations which convinced him of the superiority of the screw over paddle-wheels. The Admiralty, however, remained sceptical until Brunel arranged a celebrated tug-of-war between the sloops *Rattler* and *Alecto*, the former equipped with screw and the latter with paddles but otherwise identical. With both vessels steaming full ahead, the *Rattler* towed her rival away at the rate of 2.8 knots.

The adoption of the screw propeller set the marine engineer fresh problems. It demanded from him a new engine configuration. Also, because the rotational speed of the steam engine was then so low, it had to be geared up to the propeller shaft if the diameter and pitch of the propeller were to be kept within reasonable bounds. The engine of the *Archimedes* employed step-up gears, but Brunel decided to use chain drive on the *Great Britain*. As the diagonal crank-overhead engine had four eighty-eight-inch-diameter cylinders and indicated 1,500 h.p. it must have been some chain.

The Prince Consort performed the launching ceremony when the *Great Britain* was floated out of Patterson's Dock, Bristol, on 19 July 1843. Like her predecessor, she completed her fitting-out on the Thames. Considering the ship was not only the largest that had ever been built but combined the two novelties of iron hull and screw propulsion, her teething troubles were surprisingly few and she had made four round trips to New York before disaster struck her in September 1846.

Leaving Liverpool on the evening tide, in the middle of the night, when she was believed to be off the Isle of Man, the *Great Britain* struck and when morning dawned she was found to be high and dry on the sands of Dundrum Bay beneath the Mourne Mountains in Northern Ireland. Her master, Captain Hosken, was a most experienced navigator and such an error seemed almost incredible until it was found that, not only was there a serious mistake in the chart he was using, but the unprecedented amount of iron in the hull had affected the ship's compass. In future iron ships this was countered by mounting the compass at the head of the mast and viewing it by a periscope. This is a good example of the unforeseen perils that must beset all great engineering innovators. Because it is so easy to be wise after the event, they made engineers of Brunel's calibre an easy target for contemporary and future critics.

In fact, like the early launching mishap to the *Prince of Wales*, the results of this disaster were highly instructive. As she ran ashore, the *Great Britain* had ground her way over rocks hidden beneath the sand. No other ship in the world at that time could have withstood such punishment without breaking up. But the ten iron girders which ran from stern to stern between the hull and the bottom deck plating gave her immense logitudinal strength so that the ship survived intact and no lives were lost. After examining her, Brunel reported that: 'The ship is perfect, except that at one part the bottom is much bruised and knocked in holes in several places. But even within three feet of the damaged part there is no strain or injury whatever.' The ship was floated off in the following spring and, by dint of continuous pumping, kept afloat while she was towed to Liverpool where she was repaired and recommissioned. Her magnificent hull has since proved virtually indestructible. The ship continued in service until 1886 and was then used as a storage hulk in the Falkland Islands until 1937* when she was beached. But her hull still (1968) survives intact and there is talk of towing her away for preservation as the world's first iron screw-propelled ocean-going steamship.

In the early 1850s there was greatly increased interest in, and trade with, Australia. On so long a voyage out

---

*Such longevity is not unique. It has been estimated that a wrought-iron hull had an average life approximately three times as long as that of a steel hull, mainly due to the superior resistance to corrosion. Thus the first iron-hulled collier of 652 tons, the *John Bowes*, had a life of eighty-one years. Built at Palmer's shipyard on the Tyne at Jarrow in June 1852, as the *Villa Selgas* she was sunk on voyage in the Caribbean in 1933.

and home which, before the opening of the Suez Canal, was equal to the circumnavigation of the globe, existing steamships had to take coal at Cape Town. This involved their owners in great additional expense because, to maintain this coaling station, Welsh coal had to be shipped out from Penarth to the Cape. On the principles which had governed the design of his two previous vessels, Brunel designed a ship capable of completing this long voyage without refuelling. Th result was the mammoth *Great Eastern*, the most remarkable single feat in the whole field of Victorian engineering and the ultimate expression of that spirit of boldness and daring which uniquely distinguishes the pioneer generation of British engineers.

The hull of this Leviathan (as she was at first called) was of iron, 692 feet long with a breadth over paddle-boxes of 118 feet and a displacement of 32,000 tons. It was the prototype of all subsequent ocean liners, indeed it surpassed them in its immense strength and safety for it is doubtful whether, in similar circumstances, the *Great Eastern* would have suffered the fate which later overwhelmed the *Titanic*. On one occasion, indeed, an uncharted reef in Long Island Sound tore a hole in her outer plating eighty-five feet long and from four to five feet wide, yet the ship later docked in New York with most of her passengers blissfully unaware that anything was amiss.

With the disaster at Dundrum Bay in mind, Brunel developed the design of the *Great Britain's* double bottom into a complete and watertight double hull which extended up the sides of the ship to a height of thirty-five feet, or five feet above the deep-load line. Transverse watertight bulkheads divided the ship into ten compartments and in addition there were two longitudinal bulkheads 350 feet long and thirty-six feet apart on either side of the engine and boiler room.

Brunel decided to propel this great ship by both screw and paddles. There were reasons for this. First, it would tax engine-builders to the utmost to provide enough power to drive such a hull at the designed speed of fourteen knots, so it was desirable to provide for two engines driving two separate means of propulsion. Secondly, Brunel calculated that when the ship was running on a light draught of twenty-two feet specified by her owners, the twenty-five-feet propeller would stand five feet out of the water so that it was desirable to supplement it by paddle-wheels. The paddle engine was of the diagonal oscillating type using four cylinders and the screw engine likewise had four cylinders but they were horizontally opposed in pairs. Both were of huge size and produced together over 8,000 i.h.p., the greatest concentration of power ever kown at that time, but even so, with the low steam pressures which were then the rule, the ship was underpowered.

The *Great Eastern* was built by John Scott-Russell, a Glasgow engineer who had taken over the ship-building business of Ditchburn & Mare at Millwall, but because the river frontage was insufficient for such a huge hull, the steamship company leased the adjoining shipyard that had been vacated by David Napier's two sons. Construction was begun early in 1854, but owing to a combination of human, legal and physical difficulties which would have broken the heart of a lesser man than Brunel, the ship was not ready for launching until the autumn of 1857.

The ship was built in cradles and it was intended to push these down a slipway into the Thames so that she could be floated off on the high tide. As the dead weight of the hull at this stage was over 12,000 tons, a greater weight than man had ever attempted to move on land before, this proved a task of immense difficulty. Chains and hawsers parted, the cylinders of hydraulic presses cracked and it was not until the end of January 1858, that she was finally floated off successfully and moored at Deptford to complete her fitting out.

Brunel himself fell a victim to his last titanic brainchild. Worn out with the worry and strain of her protracted building and launching, he collapsed on the deck of his great ship on the eve of her sailing from the Thames. This maiden voyage was marred by a serious explosion on board caused by gross carelessness and, on receiving the news of this disaster, Brunel died on 15 September 1839.

As a passenger liner, the *Great Eastern* proved a disastrous economic failure. In the time it had taken to complete her, the Australian trade for which she was designed had slumped and by the time it revived the Suez Canal had been opened. As first built, this canal was too small to accept so large a ship; consequently she was confined to the Atlantic run to New York but proved far too large for the traffic then offering. But, as Brunel himself had envisaged, she came into her own as a cable layer. In 1866, at the second attempt, she laid the first successful Atlantic telegraph cable from Valentia in southern Ireland to Newfoundland. The American, Cyrus Field, was the originator of this project, but it is doubtful if his repeated efforts would ever have been crowned with success had it not been for the *Great Eastern*. Not only was she the only ship in the world capable of accommodating such a weight of cable, but her combination of screw and paddles made her uniquely manoeuvrable and therefore particularly suitable for such a delicate operation. She subsequently laid other cables between France and America and between England and Bombay.

Although she was commercially a failure, the *Great Eastern* was an engineering object lesson of the first importance. She also provided valuable experience in the methods and techniques of handling such large ships. As an example, with her original manual steering gear she became almost unmanageable in rough weather and a serious incident occurred during an Atlantic storm due to the steersmen losing control of her huge twin steering wheels. This difficulty was overcome in 1867 by the engineer John

McFarlane Gray who designed a form of steam steering gear by which the ship could be perfectly controlled by a miniature wheel on the bridge. This was one of the first powered control systems and the first to employ the principle now known as 'feed-back'.

Compared with land installations and locomotives, the working steam pressures of marine power units increased very slowly. The pressures used on Brunel's three ships, for example, were five, fifteen and twenty-five pounds per square inch respectively and this accounts for the vast bulk of their engines in proportion to the power they produced. The problem was that of the water supply to the boilers. All early marine boilers were fed with sea-water causing serious corrosion and such accumulation of salt that the boilers had to be blown down at frequent intervals. To use high pressure in such circumstances would have been extremely dangerous. On the other hand, low pressure spelt inefficiency, high coal consumption and the multiplication of boilers. Consequently bunkers and boiler room occupied much valuable cargo space. The need to provide excess boiler capacity so that one or more boilers could be isolated for blowing down while on voyage without reducing steam supply, aggravated this situation.

This difficulty accounts for the fact that, at a time when the steam locomotive was everywhere triumphant on land, steam's conquest of sail at sea was a much more protracted affair. The ultimate in sailing ship design, the clipper, did not emerge until 1849 and later examples borrowed much from the steamship in the use of iron framing (composite construction) and, in some cases, complete iron hulls, as well as iron lower masts and wire rigging the better to transmit the tremendous pull of their sails. Conversely, the steamships of the period all carried sail for three reasons, first to steady the ship in rough weather, secondly as an insurance in the event of serious engine failure and thirdly to take advantage of a fair wind to economize fuel.

At an early date engineers realized that the besetting problem of using steam power economically at sea could be solved only by creating a closed circuit, that is to say by condensing the exhaust steam and feeding the boilers with the distilled condensate. Samuel Hall patented a marine surface condenser in 1834, but the practical difficulties involved nealy ruined him. Particles of cylinder lubricant – usually tallow at this date – were carried over in the exhaust steam to clog and foul condenser tubes and boilers. This was solved by Sir William Thomson's improved design of condenser which appeared in 1856, but there was still boiler fouling and corrosion due, it was eventually discovered, to dissolved oxygen in the condensate. Not until the 1860s were the problems of the marine condenser finally solved. With that solution, working steam pressures immediately began to rise with economic results that doomed the sailing ship to eventual extinction. The marine-type water-tube boiler made its appearance and, by the end of the century, working pressures of the order of 250 pounds per square inch had become common. The higher the pressure of the steam, the more fully the principle of using it expansively could be exploited by passing the steam from one cylinder to another. First, compound expansion was introduced, then triple and finally quadruple expansion with two intermediate cylinders between the high- and low-pressure units. With the universal use of screw propulsion, the marine reciprocating steam engine now reached its definitive form. A traditionalist would call it 'a vertical direct acting' whereas we would call it simply a vertical engine with all cylinders in line, their connecting rods driving a single multiple throw crankshaft.

Such was the efficiency and economy of these marine installations that whereas the earliest steam boats borrowed their power plants from land practice, in the second half of the nineteenth century marine-type boilers and engines were increasingly adapted for power supply on land.

In association with railways on land, the steamship, by its greater size and its relative independence of the vagaries of wind and tide, brought far-reaching social and economic changes to the coastal areas of Britain. The changes in the Irish Channel ports brought about by steam packets operating in conjunction with railways has already been noticed. More profound changes accommpanied the gradual supersession of sail by steam in coastal merchant shipping. From innumerable small ports and harbours round our coasts, trade steadily ebbed away as the nineteenth century advanced, either because they were too small to accomodate larger craft or because the local trade they had served could be more economically handled by rail. Conversely, if such a port was not well served by rail, any export trade it had handled was doomed.

The use of steam on the ocean routes had effects which were even more marked by concentrating trade increasingly on a few larger ports such as Liverpool, Southampton, London and Glasgow where naturally communications favoured large-scale capital developments in the form of new deep-water quays and enclosed docks. The Caledonian Canal, engineered by Telford during the early years of the century with the object of enabling sailing ships to avoid the hazardous passage round the north of Scotland, gradually fell out of use.

One reason for the ruin of the pioneer Great Western Steamship Company was the inexplicable failure of Bristol, at that time the second port in the kingdom, to exploit the advantage which that company's leadership had given them. Not only did the Bristol Dock Company levy punitive dues on the *Great Western*, but it remained deaf to the company's appeals to widen their dock entrance to accommodate her. As a local rhymester wrote:

> *The Western an un-natural parent has,*
> *For all her beauty;*

Her mother never harboured her, and yet
  She asks for duty.
Hull, Liverpool and other ports aloud
  Cry 'Go ahead!'
A certain place that I know seems to say
  'Reverse' instead.

This betrayal by her home port combined with the lack of a sister-ship to lose the *Great Western* the Atlantic mail contract. It was awarded to Samuel Cunard whose ships used the port of Liverpool.

The *Great Britain*, though built in Bristol by a Bristol-based company, was forced by her size to trade from Liverpool from the outset, while Southampton became the home port of the *Great Western* when she was sold into other hands after the failure of her first owners. Thus Bristol failed to seize her opportunities in the age of steam.

Another port to lose its trade to Liverpool was Whitehaven. One of the six leading ports in the country at the beginning of the century, its trade began to ebb away in the 1860s because of the inability of its harbour to receive larger ships and its bad rail communications. When the leading Whitehaven ship-builders and ship-owners, the brothers Thomas and Jonathan Brocklebank, decided in 1865 to remove their business to Liverpool, the fate of Whitehaven was sealed.

In the highly competitive trade of the deep-sea merchantmen it became apparent from the dawn of the era of steam power at sea that profitability lay in increasing the carrying capacity of the ship, assuming port and cargo handling facilities could be provided to cope with this increase in an economical way. Whitehaven was an early casualty in an evolutionary process which is still continuing as modern giant tankers, container vessels and bulk carriers bear witness.

FOUR

The Engineer and the Farmer

On the long-term social consequences of the agricultural Enclosure Movement in England, opinions are much divided. Enclosure by consent had been going on ever since Tudor times, but by the Enclosure Movement we usually refer to those Parliamentary enclosures carried out under the spate of Private Acts passed from 1760 onwards and under the General Enclosure Acts of 1836, 1840 and 1845. This movement, therefore, was continued into Victoria's reign and coincides almost exactly with the first dynamic phase of the Industrial Revolution.

The social argument against enclosure is that it virtually eliminated the English peasant proprietors. Some stayed in their native village as farm labourers, but the majority migrated to the new industrial towns while others emigrated. But all lost their stake in the land and found themselves with no assets other than their skill. It can be said, therefore, that enclosures created a rootless urban proletariat.

On the other hand, it is argued that had the old pre-enclosure system of village agriculture prevailed, with its common grazing and strips of arable worked in two-or three-course rotation, it could never have supported a growing urban population. The great enclosing landlords of the eighteenth century were able to carry out comprehensive schemes of land drainage and improvement, to introduce new or improved breeds of animals and plants and to practice the

Norfolk four-course rotation. Such high farming methods and the vastly increased yields that they achieved required enclosed fields; they could never have come about under the old system of agriculture and their example was a stimulus to further enclosures. By 1850, eighteenth-century high farming methods had become general and although the agricultural industry employed less labour, the output per man and per acre was far higher and despite the rapid growth of manufactures it was still England's largest single industry. True, the fact that it lost the battle over the Corn Laws shows that its influence was waning, but their repeal did not have the immediately disastrous effect that farmers had feared. It was not until the 1870s that imports of foreign corn began to have a dire influence on English arable faming.

In view of the results achieved by 'high farming', it may seem paradoxical to say that the industry as a whole remained highly conservative, yet so it was when we consider the methods by which farming operations were carried out. Long after steam power had been successfully applied to manufacture and transport, the British farmer continued to rely solely upon the horse, while the most notable advance in agricultural mechanization was the substitution of the threshing machine for the flail.

The threshing machine was the invention of the Scottish millwright Andrew Meikle who erected his first machine on a farm in Kilbeggie in Clackmannanshire in 1787. Throughout the nineteenth century the machine was the subject of numerous patents which improved it in detail, but its basic working principle did not change. The major components at the heart of the machine, a rapidly revolving 'drum' shod with fluted metal beaters and the fixed metal screen or 'concave' which partially surrounded it and which, between them, beat out the grain from the ear, remained essentially as Meikle had designed them.

At first, the thresher was a fixed machine, erected in a corner of the barn where the corn was stored after havest and where it was driven by horses, or in some cases by water power if this was available. Being an elaborate and costly machine its rate of introduction was slow at first, but after the battle of Waterloo, when falling agricultural prices made economies in manpower essential, its use spread rapidly through the corn-growing counties of southern England. Its introduction was bitterly opposed by the farm labourers, who had relied for winter employment on the laborious and monotonous task of threshing with the flail. This opposition culminated in the agricultural riots of 1830 when many threshing machines were attacked and destroyed.

By this time, many small engineering works and iron foundries had sprung up in the rural districts and market towns of England, particularly in East Anglia, the birthplace of 'high farming'. Although these catered primarily for the farmer, their effect on farming methods was at first small, consisting mainly of the gradual substitution of cast and wrought iron for wood or stone in the construction of simple farm implements and appliances. These rural engineering works had their roots in traditional craft trades, for most of them were founded by enterprising village blacksmiths or wheelwrights who saw in the new iron age an opportunity to expand their business. With such origins, it was only to be expected that engineering methods in such shops were commonly archaic and generally backward compared with those engineering shops in the north which the textile trade had called into being. Nevertheless, it was due to their influence that that basic farm implement, the plough, was transformed in the early years of the nineteenth century from a crude construction of wood and blacksmith's ironwork into a stronger, handier and far more efficient implement constructed entirely of cast and wrought iron.

Robert Ransome, a name famous for ploughs to this day, initiated this transformation at his foundry in Ipswich in 1803 when he produced his ploughshare of chilled cast iron.

The arrow-shaped share is the most crucial part of the plough since it has to make the initial cut in the 'slice' of soil which is inverted by the mould-board that follows it. Whereas the earlier shares shod with wrought iron had worn away very rapidly due to the abrasion of the soil and so lost their points, one surface of Ransome's cast share was chilled to a glass hardness with the effect that, even when it was considerably worn, the share retained its point. This seemingly small innovation had an immense effect on the speed and efficiency of arable cultivation.

By the early 1840s, British agricultural engineers were producing improved iron-framed ploughs, cultivators, harrows and seed drills as well as a variety of barn machinery; threshing machines, chaff cutters, root pulpers and slicers and small grinding mills. But, aside from a few isolated experiments, there had been as yet no attempt to apply steam power on the farm. Barn machinery was driven by 'horse engines' or 'horse gears'. One or more horses, depending on the amount of power required, were harnessed to a long pole attached to the axis of a large crown gear. As the horses walked round and round, the power they generated was transmitted from the central crown wheel to one or more machines by means of gears, universally joined shafts, and belt drives. The round buildings that were provided to house these horse gears may still be seen on some farms, particularly in Scotland.

When the application of steam power to agriculture was first considered, however, it seemed essential that, to be of maximum benefit, the power source should be movable and not fixed. Hence the first agricultural steam engines to be produced on a commercial scale consisted of a locomotive-type boiler on four carrying wheels, the simple steam engine being mounted on the boiler top. They became known as portable engines or, more simply, as 'portables'.

Although isolated examples appeared from 1841 onwards, it was not until 1845 that the building of these portable engines began on a commercial scale at the works of Messrs Clayton & Shuttleworth of Lincoln. By 1856 they had built 2,200 engines, and by that date a number of other agricultural engineers had followed their example. The British farmer might be conservative and slow off the mark, but he readily adopted the new power when it was offered to him in suitable form.

With the coming of steam power to the farm, the threshing machine appeared in portable form on a wheeled frame and this enabled corn to be threshed straight from the rick instead of having to be carried to the barn as heretofore. This combination of steam engine and threshing machine, however, represented a capital outlay which only the largest farmers could afford, a situation which led to the birth of the threshing contractor whose modern successor operates combine harvesters on the same basis.

The early threshing contractors had to keep large teams of horses because both the engine and the threshing machine, the former especially, were extremely heavy. If the ground were at all soft it would take the combined efforts of all the horses that contractor and farmer between them could muster to drag the two machines into the rick yard and get them set up in their correct positions. It was as a partial solution to this difficulty that in 1859 Thomas Aveling made a standard Clayton & Shuttleworth portable engine self-moving by linking the engine crankshaft to the rear wheels by a single reduction gear and final chain drive. Clayton & Shuttleworth took up the idea and built their first self-moving engine in the winter of 1862–3.

This was simply an expedient to enable a heavy engine to be moved about the farm more easily. The engine still had shafts attached to the swivelling forecarriage at the front, but now only a single horse was necessary for steering purposes. Having evolved the agricultural steam engine so far, its makers were soon under pressure from farmers and contractors to take a further step by making the engine capable not merely of moving itself but of towing the threshing machine, or other implements, from farm to farm, thus dispensing with the need for horses altogether. Two-, or later three-speed gearing was introduced, while gear drive soon took the place of chains and so the agricultural traction engine was born. All the most celebrated builders of these engines were located in the eastern half of England from Leeds in the north to Rochester in the south with a concentration in East Anglia, but there were also notable builders in the southern corn-growing counties at Oxford, Basingstoke, Andover and Devizes. It became almost a point of honour for any agricultural engineer of local repute to engage in the manufacture of steam engines so that, in the heyday of the traction engine from 1870 onwards, many small rural ironworks entered the field though their combined output probably did not exceed that of any one of the great East Anglian builders such as Messrs Burrells of Thetford, for example.

The agricultural traction engine was the first type of mechanically propelled vehicle to appear on the English roads in any numbers. The notorious 'Red Flag Act' of 1865 restricted maximum speeds to four miles per hour in the country and two in towns and insisted that every engine be accompanied by not less than three men, one walking in front carrying a red flag. While this Act prevented the development of mechanical road transport in England until it was repealed in 1896, farmers and agricultural contractors willingly accepted its restrictions. The traction engine was still preferable to the horse team.

No sooner had the portable steam engine appeared on the farm than farmers and engineers began to ask whether the power of steam could not be applied in the field to the actual process of cultivation. The chief bar to the use of the steam engine for such a purpose was its great weight. If it was used merely to tow a plough or other implement it would do more harm than good, even if it did not get completely bogged down. Schemes for mechanical cultivation had been hatched in the eighteenth century, but it was not until the end of the 1840s that serious practical attempts were made in this direction. These took three forms: first, by direct haulage, using special wheels on the engine to spread its weight; secondly, by incorporating a form of engine-driven digger in the design of the engine itself, and thirdly, by using steam power to draw the implement over the field by means of cables.

The most notable exponent of the first school of thought was James Boydell whose patent wheels were first fitted to an engine in 1854 and, in the years immediately following, were used by a number of different makers, most notably by Burrells. A succession of broad flat plates were hinged to the rim of the traction engine driving wheels in such a way that, as the wheel revolved, one plate was always interposed between its rim and the ground. This was the principle of the modern track-laying vehicle, but at this date the Boydell wheel enjoyed only a very limited success. Even with this improvement, the weight of the engine was still too great to be used with success for direct haulage in the field.

James Usher of Edinburgh patented a steam plough as early as 1849 which worked with moderate success in Scotland, but although various other designs appeared subsequently and Thomas Darby of Pleshey, Essex, persevered with his Steam Digger from 1877 until the end of the century, such machines were not commercially successful. They were complicated, highly specialized and costly. Few farmers could afford such a machine solely for cultivation, while it did not appeal to contractors because of the difficulty of moving it from farm to farm.

The greatest and most successful exponent of steam cultivation by cable was John Fowler who was born at Melksham in Wiltshire in 1826 and who subsequently founded the famous firm in Leeds which bore his name. Fowler became a member of a firm of general engineers in Middlesborough where his thoughts were first turned to agriculture by the appalling consequences of the Irish famine. He pondered whether conditions in that unhappy country might be improved by systematic land drainage. He therefore developed a mole plough in which a projectile-shaped share (the 'mole') attached to the end of a deep coulter tunnelled its way through the ground drawing behind it, like beads on the necklace of a rope, a succession of wooden or unglazed clay pipes. The principle of this mole plough was sound, but to draw it directly was beyond the power of horses and this problem of traction was not easily solved. Fowler's mole plough first appeared in 1850, but it took him four years to evolve a satisfactory method of drawing it. This took the form of a double winch, each half of which could be powered at will by an ordinary farm portable engine. The rope from one winch barrel was led, via an anchored pulley, to the front of the plough, while the rope from the other passed round a pulley anchored at the opposite end of the field before being attached to the rear of the plough. This second rope served only to return the plough to its starting point preparatory to taking another working pull. Thus, with the power unit and winch set up in one corner of the field, its whole area could be drained by moving the two anchor pulleys progressively along the headlands.

When Fowler demonstrated this apparatus before the Royal Agricultural Society at Lincoln in 1854, one of the judges remarked: 'Surely this power can be applied to more general purposes.' The same idea had already occurred to Fowler, but there were serious snags. First, though such a cumbersome apparatus was acceptable for the 'once only' operation of land drainage, it was much too elaborate and slow for routine cultivation. Secondly, it would be necessary to evolve a special form of two-way plough to obviate the time-wasting idle return pull. The second problem was solved by David Grieg, an Essex farmer and a friend of Fowler, who suggested what became known as the Fowler balanced plough, the prototype of which was built at Ipswich by Ransomes early in 1856. This consisted of a see-saw frame, its pivot being a central two-wheeled axle. On this frame were mounted two sets of plough bodies, right-and left-handed. Each set could be brought into action alternately by merely tilting the frame and the need to turn the plough at the end of the furrow was thus eliminated.

With minor improvements, Fowler persisted with his double-barrelled winch and anchor pulley system for some years. In view of its tedious complications, this may seem inexplicable to us, but Fowler, marking the fate of the elaborate steam digging machine, was anxious to evolve a system of cultivation which could be powered by an ordinary farm portable engine and so brought within the financial reach of the average farmer. However, he was eventually forced to give up this idea by mounting his double winch beneath the boiler barrel of a traction engine which drove it through gears and a vertical shaft. Such an engine still required the use of an anchor pulley, so Fowler eventually abandoned it in favour of two engines, each with a single winch drum, which pulled the implement backwards and forwards between them. Such a double engine set of ploughing tackle was first demonstrated to the public at Worcester in 1863.

Although Fowler could be said to have failed in his original intention because his two-engined ploughing set was far too costly for all but the largest farmer to afford, it was, none the less, the world's first successful method of mechanical cutlivation. Fowler cable engines, and those built by other English makers, performed many formidable feats of land reclamation abroad, while at home steam ploughing, like threshing, became a job for the contractor. For, unlike the steam digger, a set of steam cable tackle could readily move on the road. At home, the use of steam ploughing engines was restricted by the decline in arable farming which set in so soon after their introduction, but they came into their own during the First World War and were not made obsolete by the farm tractor until the 1930s.

After 1870, British agricultural engineers lost the initiative to America. They were scarcely to blame for this. With English arable acres reverting to pasture under the impact of shiploads of wheat from the New World, this was inevitable. Moreover, the vast scale of arable farming in the Middle West and Canada made it possible for American engineers to apply their new mass-production techniques to the manufacture of agricultural machinery. The first of these American machines was the ingenious reaper developed by Cyrus McCormick, who had taken out his first reaper patent as early as 1834. And since McCormick's plant in Chicgo was turning out more than 4,000 of these machines a year, it was offered at a price with which no British agricultural machinery manufacturer could compete. In 1900 the McCormick reaper appeared with an ingenious knotter mechanism which bound the sheaves with twine. In the twentieth century, the Fordson tractor and the combine harvester would repeat the triumph of McCormick's reaper to make both the agricutlural steam engine and the threshing machine obsolete.

Another engineering development must be mentioned at this point because it exercised a profound influence on farming, not only in this country but throughout the world. This was the introduction of machinery which enabled fresh meat to be refrigerated and transported over great distances.

The scientific principles governing refrigeration had been demonstrated by scientists from 1755 onwards, but the first practical commercial refrigerating machine was patented in England in 1856 by James Harrison, who was born in Scotland in 1816 and had emigrated to Australia, settling at Geelong, Victoria. His first machines, which used the vapour compression principle, were built by Siebe

Brothers, afterwards Siebe, Gorman & Co., of Red Lion Court, Holborn. The first machine went to Australia and the second to Truman, Hanbury & Co., the brewers.

Other machines by English, French and German inventors soon followed until, in 1877, H. and J. Bell of Glasgow in association with the British chemist, Joseph James Coleman, patented the Bell-Coleman air refrigerating plant for ships. In 1880 the SS *Strathleven*, fitted with Bell-Coleman plant, landed the first shipload of forty tons of Australian frozen meat in the London docks. Two years later the similarly equipped SS *Dunedin* landed a first and very much larger cargo of frozen mutton and lamb from New Zealand.

This important development, while it benefited the overseas producer and helped to feed a swollen indusrial population, undoubtedly contributed to the downfall of British agriculture after 1870. With the decline of arable farming, the British farmer had pinned his faith in fat-stock. Now, when this trade, too, was threatened, it must have seemed to him that his only future lay in cow-keeping.

The repeal of the 'Red Flag Act' made road transport possible in this country and helped the makers of agricultural steam engines to survive the effects of the great farming slump. They adapted their engines for road haulage and many light 'steam tractors' and steam wagons were designed for this purpose. Furthermore, the general improvement of roads led to a great demand for heavy road rolling, a task for which the agricultural steam engine could readily be adapted. Many builders took advantage of this demand. It provided them with a market for their engines which lasted until the advent of the oil-engined roller in modern times. While France introduced the steam roller in 1860, Messrs Aveling & Porter of Rochester, a firm who became particularly famous in this field, built their first steam road roller for Liverpool Corporation in 1867.

Another lucrative but stranger field in which some firms of agricultural engineers found salvation was the making of machinery for the fairground. The original fairground ride, the roundabout or 'gallopers', was propelled by horses, but in the 1870s a Norfolk agricultural engineer named Sidney Soame of Marsham built a very small portable steam engine which he used to drive a roundabout at Aylsham Fair. By doing so he laid the foundation of a considerable industry. Soon several country engineering works were producing a variety of mechanical fairground rides: steam gallopers, steam yachts and scenic railways, while to haul these rides from town to town the sober agricultural traction engine blossomed forth as a 'showman's engine', decked out in a baroque splendour of twisted brass and columns and other decorative metalwork.

The earliest fairground rides had their own steam motive power in the form of a specialized variant of the farm portable engine known as a centre engine. As its name denotes, this formed the centre piece around which the ride was built. Later, however, the ride was driven by an electric motor supplied with current from a dynamo carried above the smokebox of the showman's engine.

The design and construction of these fairground rides was a not inconsiderable feat of mechanical ingenuity. To obtain the necessary motions, a complex mechanism was required which must be designed so that it would work reliably and safely and at the same time be capable of being quickly dismantled and reassembled many times by unskilled labour. The reliability of these curious machines and their long life is a tribute to the skill of the Victorian agricultural engineer. One of the first to profit by the example set by Sidney Soame was Frecerick Savage of the nearby St Nicholas Ironworks in Kings Lynn. He took up the manufacture of fairground machinery with such energy, skill and success that the name of Savage soon became so famous in this field that its earlier association with agricultural engineering was almost forgotten.

Nowadays, when surviving examples of the agricultural traction engine, heavy-breathing, brasswork atwinkle, lumber round the show ring, we regard them with affectionate nostalgia. Like the shire horse and the gaily painted harvest wagon they seem to have become a part of the romantic past of rural England as the townsman conceives it. But one has only to read Thomas Hardy's description of the steam threshing scene in *Tess of the D'Urbervilles* to be made aware of the true impact of steam machinery on Victorian country life. Here the steam engine and its overalled attendant are portrayed as black and alien invaders from another world, imposing upon an age-old rhythm of field labour the relentless and untiring pace of the machine. For us, the Golden Age is always yesterday. Perhaps tractor and combine harvester will one day be regarded in the same romantic light.

FIVE

The Workshop of the World

By the time Victoria came to the throne, Britain could already lay claim to the title of the workshop of the world so far as the industries of iron, engineering and textiles were concerned.

It has been estimated that by 1835 there were no less than 1,369 steam engines at work in the textile mills of Lancashire and the West Riding alone. In the coal and iron industries, too, the steam engine was omnipresent; pumping water from the pits, winding the coal and supplying blast to the iron furnaces. In this respect the steam engine was self-propagating, for the rapid introduction of steam power to manufacture and to transport could never have come about without a prodigious increase in the production of iron, a process which depended on the moulton marriage of coal and iron ore in the blast furnace.

Because they were the earliest examples of the new factory system in action upon a large scale, much has been written about conditions of life and work in and around the great steam-powered textile mills of the North since William Blake dubbed them dark and satanic. Yet they were an effect and not a cause. In order to see what was making Britain the workshop of the world it was necessary to go, not to the valleys and foothills of the Pennines, but to that heart of England which had already earned itself the name of the Black Country. Here was a spectacle far more satanic, the fiery fountainhead of the Industrial Revolution, a whole area dedicated to the production of iron.

The district surrounding the hill on which the blackened keep of Dudley Castle stands was uniquely rich in mineral resources. There was the famous Staffordshire 'Ten Yard' seam that yielded as much as 20,000 tons of coal to the acre. There were also large deposits of ironstone, the stratification being such that coal and ironstone could often be drawn from the same mineshaft. Nor was this all. Nature had supplied ample quantities of fireclay and refractory sand, the materials necessary for making blast furnaces, while Dudley Castle and Wren's Nest hills yielded the limestone that was needed as a flux in the furnaces.

This combination of mineral resources had long been known but could not be exploited until men had learnt how to smelt iron with mineral fuel instead of charcoal. Also, the fact that the district was situated on the central plateau of England at a height of from 400 to 500 feet above sea level meant that there were few rivers large enough to be used for water power and none that were suitable for transport. The coming of a network of canals to the area in the last decades of the eighteenth century solved the transport problem and, coinciding as it did with the widespread introduction of steam power and coke-smelting, initiated a process of development which the railways subsequently accelerated until, by the middle of the nineteenth century, the Black Country had become the greatest iron-producing district in Britain and one of the greatest in the world.

Today, the Black Country resembles some extinct volcano. Like many another old industrial district of Britain, only the remains of old slag tips, arid as moon mountains, now remain to remind us of furnace fires long-dead and cold. So it is difficult for us to envisage the profound impression that the district made upon the rare visitor from the outside world at a time when the landscape of England was, for the most part, still unsullied by industry. To one accustomed to the pastoral beauties of the southern counties or to the seemly and ordered calm of cathedral cities and market towns, to stand upon the walls of Dudley Castle and to gaze down at the landscape below, ringed by fires and under a perpetual pall of smoke, was like looking into the mouth of hell itself. No wonder that all the persons of sensibility tried to forget the existence of the Black Country, ignoring the fact that out of its molten travail a new and very different England was coming to birth.

James Nasmyth, one of the most eminent mechanical engineers of the nineteenth century, son of a Scottish portrait painter and himself a gifted amateur artist, visited the Black Country at this time and has set down a vivid impression of what he saw. Even this hard-headed man of the new age was fascinated and appalled.

> The Black Country is anything but picturesque [he wrote]. The earth seems to have been torn inside out. Its entrails are strewn about; nearly the entire surface of the ground is covered with cinder-heaps and mounds of scoriae. The coal, which has been drawn from below ground, is blazing on the surface. The district is crowded with iron furnaces, puddling furnaces and coal-pit engine furnaces. By day and by night the country is glowing with fire, and the smoke of the ironworks hovers over it. There is a rumbling and clanking of iron forges and rolling mills. Workmen covered with smut, and with fierce white eyes, are seen moving about amongst the glowing iron and the dull thud of forge-hammers.
>
> Amidst these flaming, smoky, clanging works, I beheld the remains of what had once been happy farmhouses, now ruined and deserted. The ground underneath them had sunk by the working out of the coal, and they were falling to pieces. They had in former

times been surrounded by clumps of trees; but only the skeletons of them remained, dilapidated, black, and lifeless. The grass had been parched and killed by the vapours of sulpherous acid thrown out by the chimneys; and every herbaceous object was of a ghastly gray – the emblem of vegetable death in its saddest aspect. Vulcan had driven out Ceres. In some places I heard a sort of chirruping sound, as of some forlorn bird haunting the ruins of the old farmsteads. But no! the chirrup was a vile delusion. It proceeded from the shrill creaking of the coal-winding chains, which were placed in small tunnels beneath the hedgeless road.

I went into some of the forges to see the workmen at their labours. There was no need of introduction; the works were open to all, for they were unsurrounded by walls. I saw the white-hot iron run out from the furnace; I saw it spun, as it were into bars and iron ribbands, with an ease and rapidity which seemed marvellous. There were also the ponderous hammers and clanking rolling-mills. I wandered from one to another without restraint. I lingered among the blast furnaces, seeing the flood of molten iron run out from time to time, and remained there until it was late. When it became dark the scene was still more impressive. The workmen within seemed to be running about amidst the flames as in a pandemonium; while around and outside the horizon was a glowing belt of fire, making even the stars look pale and feeble. . . .

Next morning, Nasmyth visited Dudley Castle and wrote:

The venerable trees struggle for existence under the destroying influence of sulphureous acid; while the grass is withered and the vegetation everywhere blighted. I sat down on an elevated part of the ruins, and looked down upon the extensive district, with its roaring and blazing furnaces, the smoke of which blackened the country as far as the eye could reach; and as I watched the decaying trees I thought of the price we had to pay for our vaunted supremacy in the manufacture of iron. We may fill our purses, but we pay a heavy price for it in the loss of picturesqueness and beauty.*

The reason why the Black Country presented such a dramatically plutonic spetacle during the first half of the nineteenth century was the extreme wastefulness of the iron-producing process. This was due partly to the seemingly inexhaustible riches of the Ten Yard seam and partly to the conservatism of the ironmasters. Very little was known about what went on in the white-hot heart of the blast furnace and the whole process was largely empirical. Consequently, having evolved a method which worked, ironmasters were reluctant to change it. The flames and smoke that belched from the open throats of the furnaces represented the waste of ninety per cent of the fuel. This loss was realized but was accepted as inevitable because it was believed that any attempt to close the mouth of the furnace must affect the quality of the iron produced. Again, furnaces were blown with cold air in the erroneous belief that there was some mysterious virtue in cold blast iron. This arose from the fact that the 'make' of iron was generally found to be better in winter than in summer. That this was due to the increased amount of oxygen in the air, and not to its lower temperature, was not appreciated.

Raw coal could not be used in the blast furnace because the amount of sulphur it contained was injurious to the iron. It had to be reduced to coke before it could be used and this operation was performed in the open by a process

* James Nasmyth, Autobiography, 1883, pp. 163–5.

very similar to that of charcoal burning. This explains Nasmyth's reference to coal 'blazing on the surface'.

These spectacular but wasteful practices continued until the mid-nineteenth century when new methods began to demonstrate their efficiency and so to overcome conservative prejudice. James Neilson proved the economy and effectiveness of using heated air in the furnace so convincingly that by 1860 hot blast iron accounted for ninety per cent of Britain's iron output. Neilson used a separately fired stove to heat the incoming air until Alfred Cowper, an associate of Friedrich Siemens, showed that this could be done far more efficiently and economically by tapping the hot furnace gases. He closed the mouth of the furnace by a 'bell' or cone which could be lowered to admit a charge, drawing off the gases whose heat was transferred to the blast by means of a form of heat exchanger known as the Cowper regenerative stove. Finally, the primitive practice of cok-ing coal in the open gave way to the enclosed 'bee-hive' coke oven. But this still wasted the by-products of the cok-ing process and it was not until 1881 that a type of oven was introduced that conserved both the gas and the tar.

Cast iron could be produced direct from the blast furnace, but the bulk of the output of the Black Country furnaces was in the form of pig iron which was subsequently remelted in puddling furnaces for conversion into wrought iron by exposing it to oxygen and so burning out the excess carbon.* The puddling furnace was charged

---

* The original puddling process invented by Henry Cort in 1784 was extremely wasteful. He used a sand bed in his furnace and as much as fifty per cent of the iron combined with the silica in the sand to form slag. In 1816, Joseph Hall, of Tipton, a Black Country ironworker, substituted for the sand a bed of iron oxide which gave off oxygen when heated. This was the process used throughout the nineteenth century and it was a vast improvement on Cort's original method.

with about five hundredweight of pig iron and it then took two hours of extremely arduous and skilled work on the part of the puddler, constantly stirring the iron with a rake or 'rabble' introduced through a hole in the furnace door, before this amount was converted into a glowing, plastic ball of wrought iron. This was a useless and spongy mass of iron and cinder which had to be pounded or 'shingled' under a heavy forging hammer to expel the cinder and form it into a bloom suitable for conversion in the rolling mill into bars. Even so, bars rolled straight from the bloom in this way were of such poor quality as to be useless. To produce good quality iron it was necessary to cut the bars into short lengths, bind them in bundles, forge them under the hammer and roll them again at least once.

The many thousands of miles of wrought-iron rails that were laid during the great age of railway construction and all the wrought-iron plates used for building bridges and ships' hulls and in boiler-making were all produced by this infinitely laborious process. Large forgings, such as the paddle-shaft of an early steamship, were built up from small bars welded together under the forge hammer, a process known as faggoting. Thus all the spectacular developments described in previous chapters would have been impossible had it not been for the titanic labours of those fierce, white-eyed men whom Nasmyth saw, leaping about like fiends amid the flames of their furnaces.

Nasmyth himself would later make a great contribution to their work. Until 1840, the type of forge hammer used in the iron trade was the tilt or helve hammer which dated back to the seventeenth century. This resembled a huge sledge hammer whose wooden shaft was pivoted in such a way that its head was raised and then let fall upon the anvil by a cogged wheel driven first by water power and later by steam. The hammer's striking rate could be altered by varying the number of cogs on its driving wheel, but its stroke could not. Hence when a very large forging was placed on the anvil, it could no longer strike an effective blow or, in the words of the smith, it became 'gagged'.

When it was planned to propel the *Great Britain* by paddle-wheels, before Brunel's decision to use a screw, it was realized that no hammer in existence could forge so large a paddle-shaft. It was to overcome this particular problem that James Nasmyth designed his steam hammer in which the hammer head was paddled vertically by an overhead cylinder and piston whose long stroke overcame the 'gagging' difficulty. Although it was never used for the purpose for which it was designed, Nasmyth's hammer was rapidly introduced in the ironworks and forges of Britain, enabling them to meet the engineer's demands for increasingly massive forgings. It was also adopted as a pile-driver and in this form it was first used in the building of Robert Stephenson's high level bridge at Newcastle.

The lot of the men who laboured long hours in the blistering heat of furnace and forge was enviable compared with those who supplied them with fuel. Coal-mining had always been a dangerous occupation, but the coming of steam power made it much more dangerous by enabling the deeper coal seams to be exploited. At such deep levels not only was there a greater risk of roof falls, but adequate ventilation became an almost insuperable problem bringing with it the risk of death by suffocation or by the explosion of fire-damp. Until the Roots blower was introduced in the 1860s, the only method of inducing ventilation in mines was the underground furnace, which was itself a hazard. Despite the invention of the miner's safety lamp by Stephenson and Davy, the fuel that powered the Industrial Revolution was won at a terrible toll of human life. At one Tyneside pit alone, the Wallsend colliery, 166 lives were lost by explosion in seventeen years. Again, in 1862 at the New Hartley colliery in the same area, the beam of the pumping engine broke, the outdoor half falling into the shaft, completely blocking it. Before the shaft could be cleared, the 204 men and boys in the pit below had perished by slow drowning. This tragedy shocked the country and led to a regulation that in future every coalmine must have two shafts, a requirement that in any event became essential when forced ventilation was introduced.

Because the steam engine was unsuitable for underground power supply, it was in the collieries that compressed air was first used as a means of power transmission. The first of such installations on record was at Govan colliery in Scotland in 1856. It consisted of a beam engine on the surface having two compressor cylinders coupled to the beam supplying air to a steam engine below ground which had been converted to run as a compressed-air motor.

Although compressed-air coal-cutting machines appeared from 1858 onwards, there was little incentive to mechanize the miner's laborious task so long as his wages remained so low. Compressed-air rock-drills were, however, used in tunnelling following their success in driving the Mont Cenis tunnel in 1857. But the principal use of compressed air in coal-mines was to drive fixed haulage engines.

In the early years of the century, William Murdock at the famous Soho Foundry in Birmingham made ingenious use of atmospheric pressure to drive individual machine tools, using small vacuum motors connected by pipe line to a central steam-driven exhauster. But this, like compressed air, was never widely used by nineteenth-century engineers as a method of power transmission and then only in small power applications.

The first and most widely used method of power transmission was hydraulic. This enabled the engineer to wield a silent and almost irresistible force greater than that of Nasmyth's steam hammer. In 1795, Joseph Bramah first demonstrated by means of his hydraulic press how, on the principle of the lever, a small hydraulic pump, acting on the large piston of the press, could be made to exert enormous pressure.

Bramah's press was first employed at the end of the eighteenth century for baling cloth, for expressing the oil from seeds and for processing hides in the tanning trade. The principle was first used in engineering in 1810 for the tensile testing of metals and from this time forward it made an increasingly important contribution to knowledge of the strength of materials. Without the use of hydraulic presses (or jacks, to use the modern term) the great spans of the Britannia and Saltash bridges could never have been raised nor the *Great Eastern* launched.

In 1846, at Newcastle, Sir William Armstrong produced the first hydraulic crane which came into widespread use for cargo-handling in Britain's docks. He followed this in 1851 with his invention of the hydraulic accumulator in which the pump raised a ram loaded with heavy weights. The return of this loaded ram supplied the power, and the device therefore provided a more convenient means of storing power than the previous method of pumping the water to a tank at the top of a tall tower to obtain sufficient head.

Later in the century, public hydraulic power supply companies were established in most of the major cities beginning with Hull in 1877. The hydraulic mains supplied power for a variety of uses, powering cranes, dock gates and passenger and goods lifts in large buildings. The introduction of hydraulic motors greatly increased the usefulness of hydraulic mains. The low-level swing bridge beside Stephenson's high-level bridge at Newcastle is still swung by its original oscillating cylinder hydraulic motor. The bridge was built in 1876 at the Elswick works of Sir William Armstrong whose pioneer installation of hydraulic cranes stood on the Tyne Quays nearby.

The chief disadvantage of these early hydraulic installations was that they were liable to be immobilized, if not damaged, by frost. Nevertheless the hydraulic supply companies resisted competition from electric power so successfully that several of them, including the London company, exist to this day. Raising the 1,000-ton bascules of Tower Bridge is probably the most spectacular current use of hydraulic power.

In the manipulation of wrought iron, powerful new hydraulic machines began to be introduced from 1860 onwards, making a contribution as great as that of the Nasmyth steam hammer. The first hydraulic forging press appeared in 1860 and in 1863 a 1,250-ton press was built by Kirkstall forge at Leeds. Plate-bending and flanging presses took the place of the arduous method of shaping plates over wooden formers with sledge hammers, while hydraulic plate-sheering, punching and riveting machines made a great contribution to the making of boilers, ship's hulls and iron bridges.

All these developments in the production and manipulation of iron could not have produced such spectacular results had it not been for the great mechanics who evolved machines for cutting the metal, thereby converting the rough forging or casting into finished components for steam engines and other machines with ever-increasing speed and accuracy. These men were the true 'back-room boys' of the Industrial Revolution. Hidden within the four walls of the machine shop, their work was less dramatic and spectacular than that of ironworks, forge and foundry with their flames and smoke. But even if it had been seen by the public, it would have been incomprehensible – to the layman anyway. Yet but for the new techniques they perfected, the rapid development of the steam locomotive in the nineteenth century could never have come about. And the steam locomotive is but one example of many.

In the eighteenth century, James Watt's steam engine remained a good idea in the mind of its inventor until John Wilkinson had evolved a machine which could bore the cylinder with sufficient accuracy. From that time forwards it became plainly apparent that engineering progress would be governed by the ability of the machine shop to translate new ideas into hardware. To take but one example, the persistence of the Watt-type beam engine with its parallel motion was largely due to the inability of the machine shop to produce slide bars of sufficient length and accuracy. As soon as this problem was solved by the aid of the planing machine, the direct-acting steam engine, with its crosshead gliding to and fro between accurately machined surfaces, began to supersede the beam engine in many of its applications.

The greatest of all the early mechanics and the pioneer of workship precision was Henry Maudslay. He died in 1831, but his influence on British engineering expertise lasted until the end of the century. Maudslay was not only an ingenious and inventive man, but a superlative engineering craftsman. He was the first to realize that workshop precision depended upon four things: accurate screw threads; true plane surfaces; absolute rigidity in all machine tools, and precise methods of measurement. The origins of the lathe, man's basic machine tool, may be traced back into pre-history, yet Maudslay's first screw-cutting lathe was the undoubted parent of the modern lathe because it was built upon these principles.

The most significant characteristic of all the engineer's machine tools is that they are self-propagating. Once the necessary accuracy has been built into them, they are capable of reproducing themselves or other, novel machines with great facility and a like accuracy. Progress in mechanical engineering, the successful translation of new ideas into three-dimensional form, is ultimately governed by this cumulative process. But someone has first got to initiate it and the story of Maudslay's first screw-cutting lathe is a good example of the way in which this was achieved. Maudslay cut the lead-screw for this lathe by laborious hand methods, taking infinite pains to ensure its accuracy. Yet once it was installed the machine could readily reproduce similar screws with uniform accuracy.

In all matters of measurement, the calibrated rule had been the engineer's sole arbiter of accuracy. Parts were 'sized' with callipers and their dimensions determined by measuring the distance between their legs against a rule. Notwithstanding the fact that a skilful and careful man could achieve surprisingly good results in this way, Maudslay realized that this method of side measurement was simply not good enough in the new machine age. Using his accurate screws, he constructed a bench micrometer capable of detecting differences of 1/10,000th of an inch. This instrument is still be to be seen in the Science Museum, South Kensington. Maudslay nicknamed it the 'Lord Chancellor' because it was used in his works as the ultimate arbiter on any question of dimensional accuracy. Maudslay thus established the principle of end measurement which has prevailed down to modern times. Only recently have optical methods superseded it for work of the highest precision.

When Sir Marc Brunel, the father of I.K. Brunel, arrived in this country from America with his designs of machines for making ships' blocks, he was directed to Henry Maudslay as being the man best qualified to make them. Maudslay first made models from the Brunel designs and later the actual machines which were installed at Portsmouth Dockyard under Brunel's supervision. Although they were designed for wood-working, these were the first examples of specialized machine tools intended for the line production of a single component in quantity. As such, they had a big influence on the future development of the engineering machine shop.*

---

* Brunel's block machines still survive, some in the South Kensington Science Museum and others in the Block Mill at Portsmouth. Maudslay's models of the machines are now in the National Maritime Museum at Greenwich.

From his small workshop in Wells Street, off Oxford Street, where Marc Brunel found him, Maudslay graduated in 1810 to Lambeth where he founded the famous firm of Maudslay, Sons and Field. For the manufacture of machinery of all kinds, but particularly for its marine steam engines, this firm was renowned for excellence of workmanship until it closed down in 1900.

No man exercised a greater influence on mechanical engineering practice throughout the nineteenth century than did Henry Maudslay. After his death, his methods were perpetuated at Lambth by his partner Joshua Field, latterly in partnership with Maudslay's two sons, Thomas Henry and Joseph, while his great-grandson, Walter Henry Maudslay, became the last managing director of the firm in 1867, dying at Falmouth in 1927. Moreover, of the great mechanical engineers of the Victorian era who extended and developed Maudslay's methods, Joseph Clement, Richard Roberts, James Nasmyth, Joseph Whitworth, James Fox and Thomas Bodmer, all but the last two were at one time pupils of Maudslay and were thus influenced directly by him.

It is significant that whereas Henry Maudslay worked all his life in London, where he was undoubtedly influenced by the precision methods of the great London makers of clocks, watches and scientific instruments, with the exception of Joseph Clement, all the great mechanics on whom his mantle fell established themselves in the Midlands and the North; Roberts, Nasmyth, Whitworth and Bodmer in Manchester and Fox in Derby. The magnet which drew them thither by promising the most lucrative outlet for their expertise was the textile industry, already highly mechanized due to the work of Crompton, Arkwright, Cartwright and Jacquard. To service the power looms, the carding and spinning machines, and to build new and improved designs, called for highly skilled mechanics. Richard Roberts particularly distinguished himself by developing Samuel Crompton's spinning mule into a self-acting machine in which one highly skilled mule spinner could supervise from 2,000 to 2,500 spindles with only unskilled assistance.

But before they could service the textile machines or build new and improved types, these great mechanics had to equip their own workshops with the necessary machine tools. These they built themselves and, having done so, found there was a ready market for them. They also found that they were equipped to build other things besides textile machinery and machine tools. Richard Roberts went into partnership with the brothers Henry and William Sharp and, as Sharp, Roberts & Co., the firm became famous as locomotive-builders. Similarly, James Nasmyth began building locomotives in addtion to his machine tools and steam hammer at his famous Bridgewater Foundry at Paticroft between Liverpool and Manchester.

James Fox originally established himself in Derby with the object of serving the cotton mills, established by Arkwright and Strutt in the Derbyshire valleys, and the Nottingham machine lace trade. But he was soon manufacturing machine tools and exporting them to Europe. The machine-tool and general engineering industry that is still to be found in the West Riding of Yorkshire had a similar origin. This can be traced back to the influence of Matthew Murray, a Newcastle engineer who was attracted to Leeds by its mechanical flax-spinning industry but whose Round Foundry soon became famous for its steam engines and machine tools. When woollen textiles replaced linen as the basic industrial product of the West Riding it was accompanied by a similar growth of engineering in all the Yorkshire woollen towns. Murray's influence, like Maudslay's, was far-reaching. One of his pupils, Benjamin Hick, later founded the firm of Hick, Hargreaves & Co., of Bolton, which became celebrated as builders of large mill engines for the cotton and woollen trades; another, Charles Todd, may be said to have established Leeds as a notable centre of the locomotive-building industry.

After Murray's death in 1835, his Round Foundry was run by a partnership known as Fenton, Murray & Jackson which became well known as locomotive-builders. Richard Peacock, co-founder of Beyer Peacock & Co., Gorton Foundry, Manchester, John Chester Craven, first Locomotive Superintendent of the London, Brighton & South Coast Railway and David Joy, inventor of the Joy radial valve gear, were among the famous locomotive engineers who were schooled at the Round Foundry at this time.

Again, William Fairbairn, a Scotsman by birth, first established himself as a millwright in Manchester. With a partner named James Lillie, he greatly improved the system of driving textile mills, substituting for the old cumbersome slow-speed power transmissions a system of high-speed transmission using light line shafting running in self-aligning bearings which was widely copied in other industries including that of engineering. Meanwhile his brother, Peter Fairbairn, founded a machine-tool-building business at the Wellington Foundry, Leeds.

So it could truly be said that those dark satanic mills in the valleys of the Pennines were the nurse of an engineering industry that, in the course of a few decades, made Britain the workshop of the world. None played a more important role in this revolution than did Joseph Whitworth. He is a key figure for two reasons. First, whereas his contemporaries launched out into general mechanical engineering, Whitworth decided to concentrate at his works at Chorlton Street, Manchester, on the production of improved engineering machine tools. Secondly, because he developed and applied more assiduously and successfully than any other engineer the lessons he had learnt from his old master, Henry Maudslay. Where Maudslay had been content with his 'Lord Chancellor' micrometer, Whitworth produced in 1855 a measuring machine capable of detecting differences of a millionth part of an inch. Where Maudslay had standardized bolt sizes and screw threads within his own works, Whitworth insisted that there should be a national standard and after much research announced in 1841 the first standard range of screw threads – the Whitworth threads by which his name became part of the everyday language of engineers. He campaigned tirelessly for greater precision in mechanical engineering, demonstrating to the Institution of Mechanical Engineers in 1857 with the aid of accurate guages that a discrepancy of one hundred thousands part of an inch could make all the difference between a driving fit and one that was loose and sloppy. 'What exact notion can any man have,' he asked scornfully, 'of such a size as "a bare sixteenth" of "a full thirty-second"?'

In view of these exacting standards, it is no wonder that Whitworth's machine tools, his lathes, his gear-cutting machines, his planing, drilling and shaping machines, acquired a reputation for accuracy that was second to none in the world. Because the Whitworth works specialized in their manufacture, he was able to supply a world demand with unrivalled efficiency, thus raising engineering standards everywhere. The contrast between his enterprise and the more generalized engineering shops established by his rivals marks a further step on the road towards greater specialization.

By concentrating upon what he considers to be the most significant achievements of any particular period the historian of engineering is apt to create a misleading picture of that period. The ideas of great innovators always encounter opposition from less enterprising mortals owing to man's innate conservatism, particularly where methods of doing a job are concerned. This conservatism is often reinforced by more practical considerations. A manufacturer may hesitate to adopt new ideas if they threaten a heavy expenditure in the replacement of obsolete plant and equipment. Similarly, his employees may resist such a change because they feel that it threatens their livelihood. So, in engineering, there is always a time-lag between the demonstration of new ideas by the more advanced practitioners and their universal adoption. Despite the infuence of his machnes, Whitworth's standards of precision took a surprisingly long time to filter down through the industry. Until the third decade of the twentieth century there were still engineering firms to be found where the use of the micrometer was rare and where terms such as a 'bare sixteenth' or a 'full thirty-second' were still current. The work of the pioneer signposts the future and seldom reflects the general level of technical competence in his day.

Nevertheless, the general progress of engineering during the first half of the nineteenth century was surprisingly rapid. William Fairbairn, delivering his Presidential address to the British Association at Manchester in 1861, summed it up in these words:

> When I first entered this city the whole of the ma-chinery was executed by hand. There were neither planing, slotting nor shaping machines; and, with the exception of very imperfect lathes and a few drills, the prepatory operations of construction were effected entirely by the hands of the workmen. Now, everything is done by machine tools with a degree of accuracy which the unaided hand could never accomplish. The automaton or self-acting machine tool has within itself an almost creative power; in fact, so great are its powers of adaptation that there is no operation of the human hand that it does not imitate.

The remarkable thing is that such a revolution should have come about in Manchester without any of the opposition, the Chartist riots and the machine-breaking, such as accompanied the introduction by the engineer of new mechanical methods of spinning and weaving. The reason was that the textile trade, like other old-established industries, was originally, to use the economist's phrase, highly labour intensive, relying upon a great number of skilled and dexterous workers, often employed in their own homes. Hence the introduction of new mechanized factory methods provoked the violent reactions with which every reader of English History is familiar. Meanwhile the engineering industry that produced such provocative machines experienced no such violent stresses and strains because it was a 'new' industry producing novel products and with no long tradition of manual skill behind it.

In the transformation of an old craft trade into a highly mechanized steam-powered factory industry the profit motive was dominant with the effect that, in human terms, the result fell far short of the Utopian dreams of the eighteenth-century pioneers. Exponents of *laissez faire* argued that the appalling social conditions in the new industrial areas were but a transitory phase and that the new system must ultimately prove to be for the good of all provided it was left to operate in freedom. The great mechanical engineers may have shared this philosophy with their mill-owning customers, but the motives which led them to evolve self-acting machine tools for their own workshops were quite different. Practical craftsmen themselves, they realized that in their new and rapidly expanding industry the work could not be done to the uniformly high standards they demanded except by the use of machines. There were limits to what the human hand, however skilful, could achieve and their demands already exceeded those limits. Moreover, among those men, mostly dispossessed countrymen and hand-weavers, who were drawn by necessity into the new textile towns, all too few possessed the necessary aptitude and skill to satisfy the Nasmyths and the Whitworths of the day. So they designed their machines, not to displace the skill of hand and eye in engineering, but to make good its deficiency. The extraordinarily rapid progress of mechanical engineering in the early Victorian period could have been achieved in no other way.

Even so, by comparison with modern methods, mechanical engineering continued to depend heavily on hand-craftsmanship, particularly in the fitting shop. It was the recognition of the advantages of standardization and the interchangeability of parts, with its demand for absolutely uniform accuracy, which would lead to the progressive extension in the engineer's workshop of the process of 'building the skill into the tool' as the century wore on. Richard Roberts was one of the pioneers of this movement and railway Locomotive Superintendents were surprised to find how readily spare parts could be fitted on locomotives built by his Manchester firm.

In the 1801 census, the population of England and Wales was nearly nine million people of whom only 2½ million lived in towns. By 1851 the urban population had swollen to ten millions out of a total of eighteen millions. If we take into account the number of mills, many still using water power, which had been established in rural areas within this fifty-year period, then the number engaged in agriculture probably remained static, if it did not actually decrease. Yet in these fifty years the population of Leeds rose from 53,000 to 172,000, while in the sixty years from 1801 to 1861 the population of Manchester rose from 84,020 to 460,028. Such was the effect, in terms of population movement and expansion, of the changeover from an agrarian to an industrial economy.

Though the average agricultural labourer's cottage may have appealed to romantic lovers of the picturesque, the living conditions it offered were far from satisfactory. Its small rooms, earth-floored below, were often overcrowded. Water supply was from a well which was usually polluted by effluent percolating through the soil from the nearby midden or cess pit. For centuries such conditions had been tolerated in rural areas, but when they became concentrated and multiplied in the thousands of jerry-built back-to-back houses and courts that sprang up around the new steam-powered factories they created a situation that could not be ignored even by the worst complacent disciple of *laissez-faire* philosophy. The workshop of the world stank to high heaven and the periodic outbreaks of cholera and typhus which took terrible toll of the working population were attributed, wrongly, to this 'miasma', as it was called.

These appalling conditions in the dense urban rookeries inspired a movement for reform whose foremost exponent was Edwin Chadwick (1800–90), Secretary to the Poor Law Commissioners. Chadwick was neither a doctor nor an engineer, but a lawyer. Nevertheless, he had a hunch that the periodic epidemics which took such frightful toll and caused so much alarm were due to the contamination of water supplies and he campaigned tirelessly for the piped supply of clean water and for better methods of sanitation and sewage disposal. Chadwick's efforts were already bearing fruit when, in 1854, his belief was confirmed empirically by the celebrated anaesthetist John Snow who discovered that an outbreak of cholera in Soho was due to the pollution by human excreta of a particular water supply pump in Broad Street. In the next decade, John Snow's discovery would be confirmed scientifically by the bacteriological researches of Louis Pasteur.

Chadwick's ideas received welcome and whole-hearted endorsement from Thomas Hawksley, an eminent civil engineer who was responsible for a pioneer water supply scheme in Nottingham and became resident engineer of that city's Trench Valley Waterworks. Asked in 1843 what had been the effect of bringing piped water to working-class homes, Hawksley replied: 'At Nottingham the increase of personal cleanliness was at first very marked indeed; it was obvious in the streets. The medical men reported that the increase of cleanliness was very great in the houses, and that there was less disease.' He concluded with these words:

My own observations and inquiry convince me that the character and habits of a working family are more depressed and deteriorated by the defects of their habitations than by the greatest pecuniary privation to which they are subject. The most cleanly and orderly female will invariably despond and relax her exertions under the influence of filth, damp and stench, and at length ceasing to make further effort, probably sink into a dirty, noisy, discontented, and perhaps gin-drinking drab – the wife of a man who has no comfort in his home, the parent of children whose home is the street or gaol. The moral and physical improvements certain to result from the introduction of water and water-closets into the houses of the working classes are far beyond the pecuniary advantages, however considerable these may under certain circumstances appear.*

Chadwick declared that three things were essential in the growing towns: a plentiful supply of clean water, pumped and distributed by pipe; a completely new system of sewers; the hygienic disposal of the sewage. To carry out these requirements, the towns would need staff to construct, operate and maintain these new works, powers to levy rates to cover their cost and powers to compel house-owners to connect their outfalls to the new sewers.

In London, sewage was discharged at low water into the tidal Thames, to be washed to and fro by successive tides, yet from this polluted source much of London's water supply was drawn without filtration. Similar conditions obtained in most provincial towns and cities. Where sewers existed they were usually of ancient construction consisting in the main, of culverted streams. Intended to carry off storm water, they were not designed to cope with solid effluent, having an inadequate gradient. Many of

* Quoted in *Human Documents of the Industrial Revolution* by E. Royston Pike.

the better-class homes had water closets of the improved design patented by Joseph Bramah in 1778 but with the addition of a water trap. But because of the inadequacy of the sewers, many householders were forbidden by law to discharge into them but were forced to make cess pits, either under or beside their houses, which were emptied at infrequent intervals. Used in this way, the water closet remained a whited sepulchre, a menace to health scarcely less great than the stinking communal privies and middens of the slums. To remedy this state of affairs in accordance with Chadwick's proposals, in London and in all the growing industrial towns, set Victorian engineers a prodigious task of public health engineering which occupied them until the end of the century.

The course of events in London, which posed the biggest problem of all, may be taken as typical. As early as 1829, James Simpson, engineer to the Chelsea Waterworks who became President of the Institution of Civil Engineers in 1853, developed a method of removing solid impurities from water by filtering it through sand. Although he did not realize it, the slime which formed on the top of the sand at his Lambeth filter beds contained organisms which fed upon impurities in the water including the disease germs. This process of natural filtration, though effective, was slow, but it was an important beginning, leading to faster methods of pressure filtration and to other more modern methods of water purification. In 1852, Simpson moved his filter beds up river from Lambeth to a site above tidal influence and in the same year the Metropolis Water Act endorsed his policy by stipulating that all water supply companies must install filtration systems by 1855 and that by 1856 their intakes from the Thames must be above Teddington.

This solved the problem of London's water supply, but the greater problem of sewage collection and disposal remained, as the foul and stinking tidal river bore eloquent witness. Long gone were the days when salmon had been caught in London's river and when Pepys and Herrick rowed on summer evenings from Westminster down to the riverside Cherry Garden in what is now London's dockland. Like the Black Country, polluted rivers were part of the price that had to be paid for a new urban and industrial civilization.

Many of the most eminent civil engineers of the day including Robert Stephenson, Thomas Hawksley and Sir William Cubitt were consulted on the problem of London's sewage. But it was Sir Joseph Bazalgette, chief engineer of the newly created (1856) Metropolitan Board of Works, who devised and supervised the great scheme that was eventually carried out and which is still in use today. He proposed a new system of sewers leading downwards towards the river, but instead of discharging into it they would be intercepted by a second system of main sewers laid parallel with it along both banks and leading to great settlement tanks at what became known as the northern and southern outfalls in the vicinity of Dagenham reach. Here the liquid effluent would be pumped out into the tideway just after high water to prevent it being carried up through London by the tide, while the solids would be loaded into bottom-dump barges to be discharged in deep water off the mouth of the river. This was the system which prevailed until recently when it was found that the dumping of solids was polluting London river through the action of the tides and modern treatment plant was installed at the two outfalls.

Water supply and sewerage systems are taken for granted today because they are unseen by the general public. We turn on our bath taps without thinking where the water comes from and flush the waste away without pausing to consider where it goes to. Yet both conveniences are the result of a prodigious feat of Victorian civil engineering. In London alone, Bazalgette's scheme involved the building of 1,300 miles of sewers and eighty-two miles of main intercepting sewers which consumed 318 million bricks. Construction coincided with the building of the Thames Embankment and the northern main intercepting sewer was laid beneath it.

In the provision of this and other improved sewerage schemes, the engineers were greatly helped by Henry Doulton. At his works at Lambeth he developed the large-scale manufacture of the 'stoneware' or vitrified pipes which were used to connect individual properties to the sewers and were a vast improvement upon the earthenware pipes of the agricultural field drain type which had previously been used for this purpose and were both fragile and pervious. Also, as we have already seen, the mechanical engineer made a vital contribution to this immense task by supplying the great number of powerful pumping engines that were required by both wtaer supply and sewage disposal schemes.

Two significant developments of the period which were destined to have an important influence on future structural engineering remain to be noted. One of these was the appearance of Portland cement which, in the form of concrete, would become the basic building material of the twentieth century. The other was the increasing use of iron in building construction. But that the two developments would ultimately combine in the reinforced or pre-stressed concrete beam, as a cast-iron beam was sometimes strengthened by wrought-iron rods, was not then foreseen.

Portland cement has no closer connection with Portland than has Canadian Cheddar cheese with Cheddar, indeed rather less so. In origin 'Portland' was merely a trade name coined by its originator, Joseph Aspdin. John Smeaton, in his account of the building of the Eddystone lighthouse, published in 1791, wrote that he had aimed 'to make a cement that would equal the best merchantable Portland Stone in solidity and durability' and this may have given Aspdin the idea for a brand name that passed into the language.

Joseph Aspdin started to manufacture his 'Patent Portland Cement' at his works in Wakefield between 1825 and 1828. Later, its manufacture was introduced to southern England through the agency of his son William Aspdin. The industry became concentrated in the area of the lower Medway valley where the necessary raw materials, clay and chalk, were most readily available. By 1852 four firms were making Portland cement at Swanscombe, Northfleet, Faversham and Rochester.

Hydraulic lime and the so-called Roman cement had been used by civil engineering since the eighteenth century, particularly for lock chambers and harbour walls requiring underwater masonry work. Portland cement was evolved from these materials and possessed much greater strength. The engineer, John Weale, in a treatise of 1850, described how 'Portland Cement as it is very absurdly called' differed in the way it was produced from hydraulic lime.

This is, in fact, nothing but an artificial hydraulic lime [he wrote], composed of the clay of the valley of the Medway mixed in certain definite proportions with the chalk of the same geological district. These materials are very finely comminuted under mills, and with water; they are allowed to deposit and then desiccated and burnt. But here begins the difference, and the most delicate part of the manufacture. Instead of merely driving off the carbonic acid gas, the calcination is pushed to a point which produces vitrification in a very considerable portion of the contents of the kiln. The lime is in fact overburnt; and it often happens that it is irregularly so. Great care is therefore required in grinding the different products of the calcination, to secure a proper quality in their times of setting. . . .

For his London sewerage system, Sir Joseph Bazalgette specified that all the sewer brickwork should be laid with Portland cement, this being the first great public work in which the material was used, the total amounting to nearly a million cubic yards of concrete. Because, as may be gathered from what Weale wrote, the cement was apt to vary in quality and strength, an assistant of Bazalgette's named John Grant carried out a series of compressive and tensile tests on sample blocks made from the cement supplied. These tests were made with equipment of Grant's own devising and were carried out throughout the period of sewer construction from 1859 to 1866. As a result, cement manufacturers greatly improved the reliability of their product.

Iron beams supported on iron columns were first used in the construction of large textile mills at the end of the eighteenth century. They were introduced solely as a precaution against fire; the exterior walls were load-bearing and the floors were supported either on segmental brick arches or flat stone slabs resting on the iron beams. The earliest known surviving building to employ this form of construction is Benyon & Marshall's flax mill at Shrewsbury* which was built by Charles Bage in 1796–7. The building is still in use and represents Bage's development of ideas imparted to him by his friend William Strutt and used in Strutt's Derby cotton mill of 1793.

Througout the first half of the nineteenth century, the form of construction used by Bage and Strutt was adopted with but minor variations for the majority of the large textile mills in the cotton and woollen districts while, as we have seen in the first chapter, iron was also used in the construction of the roofs of large railway stations. In 1840, William Fairbairn, at his shipyard at Millwall, built the first

* Now known as Jones's Maltings.

of a series of small pre-fabricated iron buildings which were subsequently dismantled and exported, but some of these, at least, depended partially for their stability on locally built masonry walls. There was no thought that iron alone might form the load-bearing frame of a building until 1850 when Joseph Paxton designed a structure which fittingly summed up the achievement of half a century. This was the first great rectilinear iron building the world had ever seen – the Crystal Palace.

SIX

High Noon in Hyde Park

A quiet green but few days since,
With cattle browsing in the shade:
And here are lines of bright arcade
                              In order raised!
A palace as for fairy-Prince,
A rare pavilion, such as man
Saw never since mankind began,
                              And built and glazed!

W.M. THACKERAY

Credit for the Great Exhibition of 1851 is primarily due to three men, its principal patron the Prince Consort, the indefatigable and tireless Henry Cole and, last but not least, Joseph Paxton who designed the unique building.

The idea of an international exhibition of arts and manufacturers was first suggested by the Royal Society of Arts, of which the Prince was President and Cole a member of the Council, following a visit by the latter to the quinquennial Exposition in Paris. A Royal Commission that included some of the most distinguished names in political and public life was appointed to put the idea into execution. This Commission in turn appointed a Building Committee to invite and consider designs for the exhibition. This consisted of the most eminent engineers and architects of the day: Stephenson, Brunel and Cubitt, Barry and Cockerell. Yet, despite this galaxy of talent, the exhibition would have been either still-born or a disastrous fiasco but for Joseph Paxton.

Having considered and rejected all the designs submitted to them, the committee produced one of their own which consisted of an immensely long brick building having a nave with two side aisles and two short transepts crowned by an iron dome 150 feet high and 200 feet in diameter. This last was the particular contribution of Brunel; it overpowered the rest of the building and made it appear as though it were being driven into the ground by an immense weight. The whole looked as absurd as an elephant riding a bicycle. It was an awful warning, which subsequent generations have, unfortunately, failed to heed, of the results of design by committee. For a design so evolved invariably represents, not the sum of individual talents, but their lowest common denominator. That such a building should be erected on the proposed site in Hyde Park not unnaturally provoked a storm of public protest so violent as to threaten to sweep away the whole idea of the exhibition.

Paxton's success story is typical of his generation. Born the son of a farmer in 1793, at the age of twenty he obtained a job at the Horticultural Society's gardens at Chiswick. There his ability attracted the Duke of Devonshire who made him head gardener at Chatworth where he displayed a talent comparable to that of Capability Brown in an earlier age. He soon became the Duke's trusted agent and major-domo, travelling with him all over Europe. The management of the great Devonshire estates brought Paxton into the railway world. He made a small fortune in railway shares and became a director of the Midland Railway.

It was in June 1850, in London that Paxton first confided his ideas for an exhibition building to John Ellis, the redoubtable chairman of the Midland Railway Company. So impressed was Ellis that he straightaway hurried Paxton off to see Henry Cole who was equally impressed and determined there and then to put the idea up to the Building Committee. But in order to do so Paxton must commit his idea to paper, and the sooner the better.

In 1849 Paxton had received at Chatsworth from British Guiana a specimen of the largest species of lily then known. He named it 'Victoria Regina' and it had since grown so extravagantly in his skilled care that he had had to design a special construction of delicate ironwork and glass to protect it. His idea for the exhibition building was based upon this lily-house, enlarged a thousand-fold. The week after his meeting with Ellis, Paxton had to preside over a Midland Railway committee at Derby. During this meeting the chairman was seen to be doodling busily on his blotting-paper. He was sketching an elevation and section of his proposed building. He took it back with him to Chatsworth where it was translated into scale drawings by the estate office staff.

Paxton's work as a landscape gardener had given him an artist's eye for line and proportion and his conception of a vast temple of iron and glass, three times the size of St Paul's Cathedral and covering nineteen acres, was essentially artistic. Yet, unlike some modern architects who, in their anxiety to escape from the cubical, stretch modern building technique to the utmost in order to achieve extravagant (in both senses of the word) geometrical forms, Paxton's design was an essentially practical exercise in structural engineering. Although it would prove surprisingly long-lived, it fulfilled, as the committee's design did not, the two essential requirements of any temporary building in being low in first cost, and capable of being erected with speed and dismantled with a like ease. It was functional and therefore timeless; in any age it would have been equally remarkable. What uniquely distinguishes it is the remarkable speed with which the building was conceived, accepted, drawn out in detail and erected. Nothing demonstrates so well the Victorian ability to get things done. Despite all our mechanical and scientific aids and knowledge, such a project today would take twice as long to complete because the vital human element is lacking; because the huge impersonal groups into which industry has coalesced, and in which the individual plays a minutely specialized role, frustrate personal responsibility and initiative.

It says much for the qualities of Stephenson and Brunel that they are once recognized and acknowledged the superiority of Paxton's proposal over their own committee's design. 'He proposed that mode and form of construction of building which appeared on first sight . . . the best adapted in every respect for the purpose for which it was intended', wrote Brunel, while 'admirable' was Stephenson's terse endorsement.

Paxton was not an engineer and his plan of the proposed building, though it determined the overall dimensions, was what we should now term a feasibility study. All the design calculations and the working drawing were made by the contractors for the building, Fox, Henderson & Co., of Smethwick, in seven weeks. Charles Fox himself* left this account of how the work was done:

It was now that I commenced the laborious work of deciding upon the proportions and strengths required

---

* Charles Fox was born in 1810 and was a pupil of John Ericsson whom he assisted with the design and construction of the locomotive *Novelty* for the Rainhill trials of 1829. He was subsequently one of the assistant engineers of the London & Birmingham Railway before founding a business of his own. Finally, in 1857, in partnership with his two sons, he founded the famous firm of consulting civil engineers which is today known as Freeman, Fox & Partners. The quotations are from his speech at a dinner given in his honour by his fellow-citizens of Derby on 21 June 1851. (See H. Clausen, *Newcomen Society Transactions* Vol. XXXII, p. 75.)

in every part of this great and novel structure, so as to ensure that perfect safety essential in a building destined to receive millions of human beings – one entirely without precedent, and where mistakes might have led to the most serious disasters. Having satisfied myself on these necessary points, I set to work and made every important drawing of the building, as it now stands, with my own hands. . . .

The drawings occupied me about eighteen hours each day for about seven weeks, and as they went from my hand Mr Henderson prepared the ironwork and other materials required in the construction of the building. As the drawings proceeded the calculations of strength were made, and as soon as a number of the important parts were prepared, such as the cast iron girders and wrought iron trusses, we invited Mr Cubitt [the Building Committee Consultant] to pay us a visit to our works at Birmingham to witness a set of experiments in proof of the correctness of these calculations.

It was found in these tests that four times the design load was required before fracture occurred. Fox then went on to describe how the building was built:

. . . . On the 26th September [1850] we were able to fix the first column into its place. From this time I took the general management of the building under my charge, and spent all my time on the works – feeling that, unless the same person who made the drawings was always present to assign each part to its proper place in the structure, it would have been impossible to finish the building in time for the opening on the 1st of May.

It was thanks to this professional zeal and leadership, coupled with a capacity for sheer hard work, which communicated itself to a labour force totalling 2,260 men at its peak, that the Crystal Palace, as *Punch* called it, was completed in seven months. In that time 4,500 tons of cast and wrought-ironwork had been erected and 6,000,000 cubic feet of timber used. The iron was erected within eighteen hours of its despatch from Fox, Henderson's foundry in Smethwick. All the glass, nearly 300,000 panes up to four feet long, was produced by Chance Brothers, also of Smethwick. There were also twenty-four miles of 'Paxton's Patent guttering'. This had three longitudinal grooves cut in it by a special machine. The glass rested in the central groove, while the other two conducted internal condensation and rain water to the vertical iron columns of the building which acted as down-spouts. The glaziers worked from special wheeled travellers running on the rain-water gutters as upon rails.

As built, the Crystal Palace differed from Paxton's original design in one important respect. Following protests about the cutting down of Hyde Park trees, a fully grown elm tree was enclosed by a lofty domed central transept, 108 feet high, which broke the long level roof line of Paxton's building. The semi-circular trusses that formed the vault of this transept were of timber – the only structural members that were not of iron. Nowadays, when public authorities and nationalized industries, ignoring protests, hack down trees and generally desecrate and disfigure the countryside with impunity, we should remember the Crystal Palace elm before we pillory our Victorian ancestors as barbarians for whom only money had meaning.

From an engineering point of view, the outstanding feature of the Crystal Palace was that it relied for lateral stability entirely on the rigid connection between vertical iron columns and horizontal beams. In this it differed from all previous iron constructions, such as the great train sheds, in which this portal bracing had been achieved either by spandrel brackets or by arched girders.

It was thus, as has been said, the first rectilinear iron building in the world. To substitute curtain walling for glass was a relatively simple step, and thus the Crystal Palace led directly to the framed building of modern type of which the Boat Store at Sheerness, designed by Colonel Thomas Greene and completed in 1860, is an outstanding pioneer example.

Like all unique ventures, the Crystal Palace attracted plenty of prophets of death and disaster. No doubt it was this unusual form of portal bracing which led to confident predictions that it would blow down in the first gale. When the building failed to oblige, despite a gale that struck while it was under construction, the prophets did not lose heart but pinned their faith in hailstorms. Even on the eve of the opening ceremony by Queen Victoria, they predicted that the salute of guns, with which it was intended to mark the occasion, would shatter every pane of glass in the building, making mincemeat of the distinguished company assembled below to welcome the Royal party.

The opening of the exhibition on 1 May 1851 was a brilliant and unforgettable occasion. In the morning, showers had rattled on the glass roof, but towards noon, the hour appointed for the Queen's arrival, the sky cleared and the spring sun burst out, setting the whole great building a-glitter against the fresh green of the park. Within, it shone down with dazzling effect upon Osler's tall glass fountain that formed a centrepiece to the Exhibition at the intersection of the nave and transepts. The royal dais had been arranged before this fountain and the red carpet leading to it was lined with ticket-holders, the women seated, their dresses of silk and satin making a vivid border of contrasting colours, and the men standing behind them. The boom of the guns and a flourish of trumpets announced the Queen's arrival. The brief interval while she retired to her robing room was followed by a storm of cheering and applause as the Queen and her Consort, with the Prince of Wales and the little Princess Victoria, entered through the great Coalbrookdale gates of bronzed cast iron and approached the dais.

Everyone was deeply moved by the splendour of the occasion, the Queen herself not least, for when it was all over she wrote in her Journal:

> The tremendous cheering, the joy expressed in every face, the vastness of the building, with all its decoration and exhibits, the sound of the organ (with 200 instruments and 600 voices, which seemed nothing) and my beloved husband, the creator of this peace festival 'uniting the industry and art of all nations upon earth', all this was indeed moving, and a day to live forever. God bless my dearest Albert, and my dear country, which has shown itself so great to-day.

The Great Exhibition continued as it had begun. It was, indeed, in retrospect, the high-water mark of nineteenth-century achievement. As that memorable summer wore on, the price of admission was progressively reduced from £1 to 1s., which gave Britain's railway system a unique opportunity to demonstrate the new mobility it offered. Thousands of country people, many of whom had never been beyond the bounds of their native parish before, now flocked to London by cheap excursion trains, drawn by the magnet of the Crystal Palace. And despite the lowering of the price of admission, when its doors finally closed on 15 October, the Exhibition had cleared a profit of £186,437 which was afterwards devoted towards founding the museums in the Exhibition Road. Paxton, Fox and Cubitt ere rewarded with knighthoods and, in addition, Paxton received an award of £5,000 from the profits.

The Crystal Palace was a great cathedral dedicated to the worship of material progress. It was as though, during that golden summer of the Exhibition, the whole nation paused to take stock and to celebrate half a century of achievement. People could contrive to forget for a while the dark roots from which the Exhibition sprang; the fire and fume of the Black Country; the ceaseless surge of the power looms and spinning mules in the great mills of the Pennines; the grinding poverty and the endless, monotonous toil; the pullulating, disease-ridden slums. For had not these things produced, as rich dung nourishes some rare exotic flower, Paxton's great glittering building and its magical contents? Suddenly, it all seemed worthwhile and there were many who shared Prince Albert's fervent belief that the Exhibition marked a turning point in man's history when friendly rivalry between nations in the useful arts would take the place of war and every sword be forged into a ploughshare. 1851 certainly marked a turning point in Britain's affairs, though not of the kind Prince Albert optimistically envisaged.

Among the engineering exhibits all the most notable achievements in civil and mechanical engineering were represented and attracted by far the greatest attention. There were models of the Britannia and Chepstow bridges and of the great suspension bridge that was in course of erection over the Dnieper at Kieff under Charles Vignoles' direction with ironwork supplied by Fox, Henderson & Co. Sir Samuel Brown, the pioneer of suspension bridges, exhibited a model of his Brighton chain pier together with a specimen of his chain cable, which he had patented as long ago as 1810 and which Thomas Telford had adopted for his suspension bridges at Conway and the Menai. The civil engineers, Robert and Alan Stevenson (grandfather and uncle respectively of Robert Louis Stevenson), showed models of their lighthouses at the Bell Rock and Skerryvore. There was a huge model of the new docks at Liverpool.

In the railway section there were the latest patent signals and brakes and an impressive array of resplendent steam locomotives which included the giant *Liverpool*, built to the design of Thomas Crampton by Bury, Curtis & Kennedy of that city for the Southern Division of the London & North Western Railway. Yet even the *Liverpool* was eclipsed by Daniel Gooch's eight-feet single express locomotive *Lord of the Isles* for the broad-gauge Great Western Railway. In gleaming green livery and glistening with highly polished metal, this latest masterpiece from Swindon towered over its narrow-gauge rivals and must have warmed the hearts of Brunel and his broad-gauge supporters.

Nearby there were marine engines by Maudslay, James Watt & Co., of the historic Soho Foundry and the Rennie Brothers. There was James Nasmyth's steam hammer, a model of one of William Armstrong's hydraulic cranes and the actual hydraulic press which had raised the tubes of the Britannia Bridge. There was a mechanical printing press, a self-acting spinning mule, a huge Jacquard loom and a bewildering assembly of machine tools, including a number by Whitworth. Many of these machines were shown in motion, steam being supplied from a special boiler house, equipped by Armstrong's Elswick Works, which stood outside the building.

Today, browsing through old engravings and photographs of the Exhibition, one is immediately struck by the extreme contrast between the engineering exhibits, which shared with the building which housed them a clean, functional beauty of line and proportion unmarred by any excess, and the galleries devoted to arts and manufactures. For the latter, for the most part, resemble some vast lumber room where a jetsam of miscellaneous objects, some pathetic, some ludicrous, but all representing a prodigal misuse of labour and ingenuity, lie gathering dust. It is the great paradox of the Victorian era that the same people who set a steam-powered girdle round the earth should carve for their Queen's palace at Osborne a garden seat out of a block of coal, admire a statue made of zinc or wear cuffs spun and woven from poodle fur.

The Victorians allowed their admiration for painstaking work and ingenuity to override their aesthetic judgement. The more eloquently an object of applied art revealed these two qualities the more highly it was prized. Hence the Exhibition galleries were filled with objects fashioned with infinite pains from some improbable material or in which some base substance had been ingeniously disguised to simulate a rarer. Furniture, china, glass, plate and textiles were generally grossly overburdened with elaborate ornament in a bewildering variety of styles, often incongruously blended. And if the ornamentation was produced, not by misapplied hand-craftsmanship, but by some novel mechanical process, the Victorians were no whit abashed. For above all things the age revered the boundless capacity of its engineers, and the machine-made object was admired for the ingenuity of the mechanical process which could reproduce with such facility the more costly hand-made article.

The only historical precedent for this was the obsession of wealthy Englishmen in the sixteen and early seventeenth centuries with the virtuosity of the wood-turning lathe which resulted in the gross bulbosities of some of the ugliest furniture ever made. Now there was a similar but more general lapse, for not only had the Industrial Revolution produced a larger and wealthier middle class, but many machines and processes more elaborate than the simple wood-turning lathe with which to gratify the undiscriminating taste of that class. The Great Exhibition reveals most clearly the ironical paradox that in a time of anarchic aesthetic eclecticism the engineer stood almost alone in maintaining the tradition of functional honesty in design and workmanship.

Probably no generation has taken themselves more seriously than the mid-Victorians. They took their cue from Prince Albert, that earnest, humourless, admirable and devoted Teuton. Posterity has found the spectacle of some grave tall-hatted figure studying an object of outrageous aesthetic absurdity irresistibly amusing. And because the libertine habitually equates moral earnestness with hypocrisy, ridicule has been liberally tinged with malice. No other age has been so unmercifully caricatured whereas, by its world-changing achievements in engineering alone, none has a stronger title to be treated as seriously as it took itself. Today we have belatedly recognized this and are making amends. Yet in contemporary literature, the occasional snide or patronizing remark reveals that bad old habits are hard to break.

The outbreak of the Crimean War a mere two years after the Great Exhibition closed its doors put an end to the optimistic notion that the art of war had been superseded by the arts of manufacture. Appalling blunders in the conduct of that war demonstrated that military methods that had proved themselves at Waterloo were about as appropriate as a medieval tournament in the new age of steam power and machines.

With this exception, Britain settled down to what is now generally regarded with nostalgia as the golden afternoon of the Victorian age. Apart from those faraway stalwarts who were extending the borders of an Empire upon which the sun reputedly never set, the Queen's subjects enjoyed decades of undisturbed peace and with in an unexampled prosperity. Yet, unlike the bracing atmosphere of the Victorian morning, this afternoon was sultry and oppressive and while there were those who basked complacently in its sunshine, there were others more perceptive whose ears caught the rumour of distant thunder presaging a troubled evening and a stormy night.

The pioneer generation of engineers were either growing old and seeking retirement or dying untimely, burned out by their own energies, worn out by the pressures of the new world they had helped to create. Isambard Brunel, Robert Stephenson and Joseph Locke died within months of each other in 1859–60. They were the last inheritors of eighteenth-century certitude in the powers of human reason as the instrument of man's emancipation through material progress. This gave them their self-confidence and daring; gave them an ability to bestride their profession and the courage to confront any task presented to them, however formidable, that had won them immense popular prestige and from their subordinates devoted loyalty. When Robert Stephenson died in October 1859, the whole nation mourned him. By special permission of the Queen, his funeral cortege passed through Hyde Park on its way to Westminster Abbey where his body was buried beside Thomas Telford, and the whole route was lined by silent crowds. In his home country, all shipping lay silent on Tyne, Wear and Tees, all work ceased in the towns and flags flew at half-mast. In Newcastle, the 1,500 employees of Robert Stephenson & Co. marched through silent streets to a memorial service. It was as though a king had died, but where was his successor?

Never again would a British engineer command so much esteem and affection; never again would the profession stand so high. To some extent this was inevitable, for as knowledge rapidly accumulated, so the profession ramified. Increasingly, teams of more or less anonymous experts became responsible for new developments in engineering rather than a single brilliant individual. In this process, with but few notable exceptions such as Parsons or Ferranti, the engineer became a smaller, more specialized species. No longer bestriding his profession with such easy assurance, he became the servant instead of the master of the boardroom and the accountant.

But this was not all. The public began to lose confidence in the engineer so that he began to lose confidence in himself. That mood of uncritical reverence for material progress which had made the engineer the hero of the hour did not long outlive the Great Exhibition. The golden calf of the new Utopia, of man's emancipation by machine, was worshipped at the Crystal Palace for the last time. A mood of doubt and disillusionment began, slowly at first, to undermine the old, easy confidence. Just as the engineer had shrunk the physical world and brought about a major social revolution by the power of steam, so it appeared as if the scientists were deliberately undermining the whole structure of thought and belief, of faith and morals which had hitherto been taken for granted. Charles Darwin published his theory of the origin of species in 1859. The very earth, which had hitherto seemed so firm-set, now shook beneath men's feet. Suddenly all became unsure and bewildering. Utopia looked as remote as ever. Was this really the right way, the path that man was pre-destined to follow, or had he taken a wrong turning? The older generation looked back wistfully to the England before railways as to some lost garden of Eden. In recollection it seemed an infinitely beautiful and peaceful country of green fields and small quiet towns where there was no smoke or thunder of machines, where there were no cracks in the old hierarchical order and God was still in his heaven. Such a pre-occupation with the past as a refuge and solace from the harsh realities of the present would become a common symptom of the malaise of industrial man. But it was then novel. Previous generations had looked forward to their Golden Age.

But it was not only among the old that symptoms of doubt appeared. The attitude of the artist towards the works of the engineer underwent a significant change.

O Satan, my youngest born, art thou not Prince of the
    Starry Hosts
And of the wheels of Heaven, to turn the Mills day &
    night?
Art thou not Newton's Pantocrator, weaving the Woof
    of Locke?
To Mortals thy Mills seem everything. . . .

125

A Scheme of Human conduct invisible and incomprehensible.
Get to thy labours at the Mills & leave me to my wrath. . . .
Anger me not! thou canst not drive the Harrow in pity's paths:
Thy work is Eternal Death with Mills & Ovens & Cauldrons. . . .'

Ever since William Blake had written these lines in his poem *Milton*, the artists of the Romantic Movement had been fascinated by the works of the new industrial age. Like James Nasmyth, himself the son of an artist, they had been at once enthralled and appalled by the implications of fierce power and lofty ambition they so dramatically conveyed. The infant industrial England they knew was full of vividly dramatic contrasts, for the boundaries between the old world and the new were still sharply defined. The new regime had not yet gained the power to blur those boundaries by spreading its desolation far over the landscape. A many-windowed mill, gaslit and pulsing with power, a fuming, flaming ironworks or a towering railway viaduct would suddenly appear in the midst of a paradisial landscape. The artist found the extremity of these contrasts 'awful' or 'sublime' and he would paint such a landscape, not by the light of the sun, but in bold chiaroscuro against the sky-reflected flare of furnace light as though summoned prematurely from age-long repose by the false promise of some unnatural daybreak. In such visions as these, the parallel between industrial man and Milton's Satan, first drawn by Blake, was implicit. It was more explicitly stated by John Martin, that eccentric painter of apocalypse, who used images drawn from the new industrialism in his illustrations for *Paradise Lost*.

But as the manifestations of industrialism grew more familiar and widespread, and less crude and violent, though no less ugly, as the engineer's methods became more efficient and therefore less spectacular, so they ceased to enthrall the arist. Satan breathing fire and sulphurous smoke had been an awful figure, fit subject for art; Satan disguised in frock coat and tall hat was not. Before this seemingly respectable yet more sinister and subtly destructive fiend and his followers the artist recoiled in revulsion, filled with dire foreboding.

> They are most rational and yet insane;
>   An outward madness not to be controlled;
> A perfect reason in the central brain,
>   Which has no power, but sitteth wan and cold,
> And sees the madness, and foresees as plainly
>   The ruin in its path, and trieth vainly
> To cheat itself refusing to behold.*

Asked William Morris:

> Who now shall lead us, what God shall heed us
>   As we lie in the hell our hands have won?
> For us are no rulers but fools and befoolers,
>   The great are fallen, the wise men gone. . . .†

Many sought refuge from a harsh present in an imaginary past of heroic myth, medieval legend or Celtic twilight, a process of withdrawal that reached its logical conclusion in the decadence of the 1890s. Yet even here it proved impossible for long to exorcise the disquieting sense that a once familiar world had suddenly become estranged and hostile.

* James Thompson, *The City of Dreadful Night*, 1874.
† William Morris, *The Voice of Toil*, 1884.

Perhaps this change of mood is reflected most clearly in two of Tennyson's poems, *Locksley Hall*, published in 1842, and *Locksley Hall Sixty Years After*, which he published in his old age in 1886. The first poem ends in an exultant paean to material progress:

> Not in vain the distance beacons. Forward, forward let
>   us range,
> Let the great world spin forever down the ringing grooves
>   of change.
> Thro' the shadow of the globe we sweep into the younger
>   day:
> Better fifty years of Europe than a cylce of Cathay.

The second sounds a note of extreme pessimism and disillusion:

> Gone the cry of 'Forward, Forward,' lost within a
>   growing gloom;
> Lost, or only heard in silence from the silence of a
>   tomb.
> Half the marvels of my morning, triumphs over time
>   and space,
> Staled by frequence, shrunk by usage into commonest
>   common-place!
> 'Forward' rang the voices then, and of many mine was
>   one.
> Let us hush this cry of 'Forward' till ten thousand years
>   have gone. . . .*

To many thinking people, the machine-dominated society that Samuel Butler portrayed in his *Erewhon* of 1872

---

* The career of H.G. Wells from ardent progressivism to the ultimate despair of *Mind at the End of its Tether* affords an interesting later parallel.

seemed a more likely destination than the bright Utopia envisaged by eighteenth-century liberalism. So engineer and artist went their separate ways to the detriment of both, with only the architect attempting with small success to bridge the ever-widening gulf between them. In the process, engineering, which had been hailed as the foremost of the Useful Arts, would come to be referred to as a branch of technology, a horrible but necessary word.

Today, despite the fact that the menace of the atomic bomb has confirmed the most pessimistic forebodings of our Victorian ancestors, we take material progress for granted; a process is inexorable as the incoming tide. It has even received canonical blessing as part of man's fore-ordained evolutionary development but without any accompanying sense of individual responsibility for it. So-called 'advanced' countries are those capable of wielding the greater power of destruction. The rest are known as 'backward'. Judged from this modern standpoint, the misgivings of the later nineteenth century must be adjudged a failure of nerve. It was one reason why Britain so speedily lost her proud position as the workshop of the world although its effects cannot be measured. But there were other reasons, complex in their interactions, but more mundane and therefore more ponderable.

Until the mid-century, the distinction between the useful arts and science had remained clear-cut. The early technical triumphs of the Industrial Revolution were, and were seen to be, so many victories for the engineer. This was one reason for his unique national prestige. The days when novel engineering achievement would be hailed by press and public as 'a new scientific breakthrough' were still far off. If engineering was a useful art, with the emphasis on art, science was a department of philosophy – 'natural philosophy' – concerned with the discovery of natural laws and the formulation of theories to explain their working. From James Watt onwards, engineers had drawn freely upon scientific knowledge and method – indeed their successes would otherwise have been impossible – but they mistrusted scientific theories and formulae, regarding them as no substitute for their practical experience and using them as a check against results arrived at by their own skill and judgement and by their own empirical tests. All science was then pure and the phrase 'applied science' had no meaning. As late as the 1860s, one of the most eminent civil engineers of the day could contemptuously dismiss certain scientific formulae on the proportions of retaining walls as having 'the same practical value as the weather forecasts for the year in Old Moore's Almanack'.

When we recall that fruitful eighteenth-century collaboration between engineers, industrialists and scientists exemplified by the Lunar Society and similar bodies in London and Edinburgh; when we realize that the subsequent period did not lack British scientists as brilliant in their own fields as were their engineer contemporaries, this contempt for scientific knowledge seems unaccountable. It was probably due to an invidious comparison that was drawn between Britain and France.

In the last decades of the eighteenth century, France had led the world in scientific knowledge and in the application of scientific theory to engineering. Yet British engineers had, by means that were largely empirical, far outstripped France in technical innovation within the space of a few decades. That the reason was largely political and economic; that the British engineer and industrialist operated in conditions of freedom unknown in France and enjoyed a unique export trade was not recognized, and Britain's proud position as the workshop of the world was attributed to some mysterious, innate engineering genius peculiar to the British. While others wasted their time poring over calculations, theories and formulae, the British engineer, by sheer talent and inspiration, had produced a better job. This myth would have unfortunate effects.

The fact that when British scientists did venture to pronounce on engineering matters they so frequently made fools of themselves appeared to confirm the myth. Even the great Sir Humphrey Davy made a woefully bad impression by his arrogant and ill-natured refusal to believe that an unlettered colliery engine-wright like George Stephenson could possibly have invented a miner's safety lamp as good as his own. The pompous and totally erroneous pronouncements on railway matters and on the impossibility of ocean steam navigation by Dr Dionysius Lardner in the 1830s were only equalled by those of Sir George Airey, the Astronomer Royal, later in the century. Having announced in a pamphlet that the Crystal Palace would certainly blow down, Airey proceeded to damn the Atlantic telegraph cable project, saying that: '. . . . It was a mathematical impossibility to submerge a cable at so great a depth. . . . If it were possible, no signals could be transmitted through so great a length.' A characteristic of such men was an unpuncturable self-esteem which led them to persist in their folly no matter how many times the engineer proved them wrong. Unfortunately, Airey's final fatuity would have disastrous consequences, as we shall see in a later chapter.

Lardner and Airey may not fairly represent an age which produced Michael Faraday, but such repeated gaffes undoubtedly fed the engineer's sense of superiority and his contempt for scientific knowledge. This led to his refusal to admit the importance of scientific education and training until it was far too late. The Institution of Civil Engineers was the first to introduce entrance examinations as late as 1897, while the Institution of Mechanical Engineers followed in 1912. The delay might not have been so long had it not been for the untimely death of the Prince Consort, the most enthusiastic Royal patron of the useful arts and sciences since Charles II.

Theoretical training cannot alone make a brilliant engineer, but at a time when increasing knowledge was making the profession of engineering more of an applied science and less of an intuitive art, the lack of it could handicap a good one and have a disastrous effect on the performance of the profession as a whole. That other countries quickly recognized the importance of technical education at a time when Britain continued to rely on the myth of innate superiority is another reason why she so soon lost her pre-eminence as an industrial nation. There were three other reasons for this decline, reasons closely associated with the first and with each other: greater religious toleration, the growth of snobbery and the cult of amateurism.

The part played by Non-conformists in the first dynamic phase of Britain's Industrial Revolution was immense but, apart from Arthur Raistrick's excellent study* of the early activities of the Society of Friends, this fascinating subject has too seldom been emphasized and never deeply explored. Non-conformists were barred from the Universities until 1870; in the eighteenth century the learned professions were closed to them, while, because they were prevented from trading in the older corporate towns and cities, they tended to set up their businesses in the Midlands and the North.

Non-conformity was then punishable by punitive fines and imprisonment, but one of the lessons of history is that such persecution of a minority invariably has an opposite effect from that intended. A refusal to conform in such circumstances requires great determination, individuality and strength of mind. Persecution fosters these qualities in men as fire tempers steel; it acts as a challenge to their initiative and forces them to cooperate one with another, sustained by the certitude of a strong, shared faith. That the Industrial Revolution originated in the Midlands and the North is usually explained in terms of natural resources.

*Quakers in Science and Industry. See Bibliography.

That it was also due to the recognition and successful exploitation of those resources by men of the highest calibre and initiative who had been virtually outlawed by the Establishment is seldom recognized.

Of all the trite sayings, 'nothing succeeds like success' is perhaps the furthest from the truth. Just as surely as persecution and adversity temper character so easy success softens it. In the Victorian age the successors to the pioneers reaped a rich harvest from the seeds which their forebears had sown. They became immensely wealthy, while the effect of a growing religious toleration was that the Establishment doors which had hitherto been firmly closed to them now opened one by one, their stature diminishing with their growing conformity as each threshhold was crossed. At length, only the door to the innermost sanctuary, that of the landed gentry, remained firmly barred. Riches fostered snobbery. Wealthy manufacturers made themselves ridiculous by their efforts to ape the gentry, but they beat against this last locked door in vain. They realized ruefully that entrance was one thing that their money could not buy.

But where they failed, their sons or grandsons might succeed now that the public schools and universities were open to them to provide the necessary polish. This polish included assuming the prevailing attitude of the gentry which was to regard as beyond the pale everyone outside a charmed circle embracing the armed services, the learned professions and the established Church. 'Trade' was for them a dirty word which damned the manufacturer and the engineer along with him. This attitude not only robbed the engineering profession of a great deal of the most promising human material but it is also the reason why engineering education was excluded for so long from the curricula of English public schools and universities. It was not considered a suitable vocation for gentlemen.

Thus, although the Victorian educational system may have produced great Empire-builders, soldiers and administrators, it failed to produce great engineers. Because, in the minds of gentlemen, business professionalism was tainted with 'trade', the system produced amateurs on the model defined by Thomas Arnold in which high principle and gentlemanly conduct rated higher than intellectual ability and a classical education was considered essential. There may be some truth in the belief that only such a system can produce the best, because the most enlightened and impartial, administrators, but its effect upon British engineering was baneful. It produced a climate in politics and finance that was unsympathetic to engineering progress because its problems were not understood. It fostered colonial expansion, not only with men but with money, at the expense of development at home. Finally, this cult of amateurism, while it emasculated the engineering profession, paradoxically subscribed to the myth of British engineering infallibility. The governing *élite* complacently refused to believe that any other nation was capable of overtaking the overwhelming industrial lead that a pioneer generation of engineers had won.

But for this myth, it would surely have been obvious that no one nation could hope, in a world of expanding commerce, to retain an easy monopoly of new ideas. The seeds of Britain's Industrial Revolution had been sown world-wide, not only through the object lessons of her exported machines, but by her expatriate engineers. It was the height of folly to suppose that they would fail to germinate.

Mention has been made in an earlier chapter of the French ironworks at Charenton founded by Aaron Manby which had a seminal influence upon French industrial development. In 1847, George Stephenson warned the Institution of Mechanical Engineers in his Presidential address that: 'it was not unlikely some of the Continental talented men might take part of the business of this country', but his warning was not heeded. In the year of the Great Exhibition, a brilliant young engineer named Charles Brown on the staff of Maudslay, Sons and Field accepted an invitation to join the small Swiss engineering firm of Sulzer Brothers at Winterthur. Not only was Brown largely responsible for building Sulzers into a large engineering business of international repute, but he also founded the Swiss Locomotive Company in 1871, and established an electrical department at the Oerlikon Works in 1884. Moreover, with his advice and help, his two sons, Charles and Sydney Brown, founded the famous Swiss firm of Brown, Boveri & Co. Thus one British engineer, schooled in Maudslay's methods, played a major part in establishing the Swiss heavy engineering industry.* The brothers Krupp received their training at the works of Messrs Smith, Beacock & Tannett, machine tool makers of Leeds, who had acquired in 1843 the Round Foundry made famous by Matthew Murray. On their return to their own country, the brothers founded the famous Krupp Works at Essen which became the core of Germany's industrial might.

But the first threat to Britain's industrial supremacy came, not from Europe, but from America. Following the War of Independence in 1785, the British Government placed an embargo on the export of any tool, machine or engine to America and did its utmost to prevent skilled engineers and craftsmen from emigrating thither. By stimulating native effort, no move could have been better calculated to be self-defeating. Two of the English craftsmen who evaded the ban were William Crompton and Samuel Slater who carried the secrets of Arkwright and Strutt to America and founded the textile industry in the Merrimac valley in New England, which, like the parent industry in old England, proved a wet-nurse of mechanical engineering talent.

* I am indebted to Mr Hugh Clausen for drawing my attention to this information concerning Charles Brown.

135

The general-purpose machine tools developed by Joseph Whitworth remained the stock equipment of the average British engineering machine shop until the end of the century. Craftsman-built, they had superb lasting qualities, while with ample supplies of relatively low-paid labour available there was no incentive to replace them by more labour-saving equipment or to carry any further the process of building the skill into the tool with its attendant advantages of higher output and interchangeability of machined parts. In America, on the other hand, where employers had to offer high wages in order to hold skilled men, and where there was a high labour turnover of unskilled immigrant labour whose one concern was to make enough money to go west and set up on their own, conditions favoured further mechanization and mass production. American progress was so rapid that in 1851 the New England engineering firm of Robbins & Lawrence exhibited at the Crystal Palace a number of rifles made by machine methods so accurate that their component parts were readily interchangeable. This exhibit attracted far less attention than Hiram Power's voluptuous nude statue 'The Greek Slave' which stole the show in the American Section, but its significance was not lost on the discerning. It won Robbins & Lawrence a medal; further, it took British engineers, among them Joseph Whitworth himself, across the Atlantic to study American production methods. As a result of this visit, the new British Government armoury at Enfield was equipped with American machine tools. These included milling machines which no engineer in England had ever seen before.

To a generation less complacent, the lesson of this episode would have been plain. Yet as late as 1914 an English locomotive engineer could write: 'American competition, owing to her entirely different, one might say indifferent, standards will never threaten this country's locomotive exports until she can build with something like accuracy and finish.' This reveals the failure of the British engineer, building machines of superb quality and durability, to understand the American engineer's philosophy. 'Why lavish so much material and labour on a machine which progress will make obsolete long before it is worn out?' the American would ask. This was the United States version of the gospel of material progress, a doctrine of 'planned obsolescence' which would dominate the world in the twentieth century. In that insignificant American exhibit at the Crystal Palace, the writing was on the wall for the workshop of the world.

For all these closely interlinked and complex reasons, the story of Victorian engineering in the period after 1860 is very different from that of the heroic years that went before. Increasingly, we find British engineers adopting and developing new inventions initiated in other countries. Sometimes they grasped the technical leadership but failed to exploit it for lack of financial support from a country that had lost faith in their powers. Sometimes a promising innovation originating in this county would stagnate while development continued abroad, eventually returning to England in a vastly different form. The history of Portland cement illustrates this last point. Its manufacture originated in this country, as we have seen, but after 1860 the initiative passed first to Germany and then to France, its strength being progressively increased. Finally, it was the French who first successfully exploited the idea of reinforcing concrete with steel. Reinforced concrete was introduced to this country by French licensees at the end of the century and at first made slow headway in structural engineering owing to the doubts of conservative British engineers about its lasting properties.

Had they but known it, the sponsors of the Great Exhibition had organized in Paxton's glittering palace of iron and glass, not the celebration of the dawn of a new era, but the grand finale of an age that would rapidly pass away. Things would never be quite the same again.

137

SEVEN

The Age of Steel

At the beginning of the nineteenth century Britain's output of pig iron had been 258,000 tons. By the year of the Great Exhibition this figure had risen to 2.7 million tons of which by far the greater proportion was converted into wrought iron. As we have seen, this was a slow and laborious process in which the quality of the metal produced was determined by the skill and judgement of the individual ironworker. The object of the puddling furnace and of the arduous stirring or 'rabbling' process carried out by the puddler was to burn out the excess carbon from the iron that rendered the metal brittle. By the 1850s this operation had become what we should now term a serious bottle-neck, limiting production of the basic material on which the progress of the Industrial Revolution depended. This forced engineers to consider whether there might not be some quicker and more reliable method of removing excess carbon from the molton iron.

James Nasmyth conceived the idea of substituting for the puddler's 'rabbling bar' a jet of steam, blown through the pool of molten metal in the puddling furnace. He argued that the oxygen in the steam would combine with the carbon in the iron to produce the desired effect. Nasmyth patented this process in May 1854, and he tells us that it was tried by several iron-manufacturers with promising results. However, though it was quicker and far less laborious, it was unreliable, depending on the very careful regulation of the steam by the puddler.

In 1856 Nasmyth was present at a meeting of the British Association at Cheltenham when Henry Bessemer read his paper 'On the Manufacture of Iron and Steel without Fuel'. 'I . . . listened to his statement with mingled feelings of regret and enthusiasm,' wrote Nasmyth, 'of regret because I had been so clearly anticipated and excelled in my performances; and of enthusiasm – because I could not but admire and honour the genius who had given so great an invention to the mechanical world.* But if Nasmyth recognized the importance of Bessemer's invention, the majority of those present did not and his historic paper was not printed by the Association. Nevertheless the paper, with its misleading title, marked the beginning, albeit a faltering one, of a new phase in the Industrial Revolution.

Sir Henry Bessemer (he was knighted in 1879) was the son of a French refugee and was born as Charlton, Hertfordshire, in 1813. He was neither a metallurgist nor an ironmaster but a professional inventor, a rare phenomenon in the England of the 1850s. He had invented a new type of gun with a rifled barrel for use in the Crimean War and, in order to make the barrel strong enough, he required a metal with similar properties to steel but capable of being cast in a mould. It was this need that had led Bessemer to undertake a series of small-scale and purely empirical experiments at Baxter House, St Pancras, in December, 1854. After several expedients had failed, Bessemer tried the effect of blowing hot air through the firebridge of his experimental reverberatory furnace in which a small amount of molten pig iron had collected. The effect of the oxygen in the air uniting with the carbon in the iron was instantaneous and spectacular. The temperature of the melt rose and the pig iron 'boiled' violently, throwing

* James Nasmyth, *Autobiography*, p. 367.

out burning carbon in a shower of sparks.* Some say that
Bessemer was induced to make this experiment by his
knowledge of Nasmyth's patent; others that his brother-in-
law, William Allen, who assisted in the experiments and
had just returned from America, communicated to him
the knowledge of the American William Kelly's similar
'pneumatic method'.†

Opinions also differ as to the quality of the metal that
was left in the furnace when the experiment was over.
Some insist that it can only have been useless burnt iron,
but had it not been more promising than this, Bessemer
would scarcely have pursued the idea further by inventing
his famous converter. This was a cylindrical steel vessel
lined with ganister and mounted in trunnions so that it
could be tilted. At one end were a series of air-inlet nozzles
or 'tuyeres' and at the other an open mouth. The converter
could be tilted to receive through its mouth a charge of
molten metal from the blast furnace, then raised to the
vertical 'blowing' position when air was blown through the
metal, and finally tilted once more to allow the processed
metal to be run out. A big Bessemer converter blowing
gave a very fair imitation of Vesuvius in eruption and a
previous generation would not have hesitated to call the
spectacle 'awful' or 'sublime'.

Bessemer called the metal which resulted from this
process 'steel'. This was confusing as in its characteristics it
bore no resemblance to the costly steel hitherto produced

* Had he but known it, in this experiment Bessemer had come very
close to the open-hearth method of steel-making referred to later in this
chapter.
† Kelly's discovery antedated Bessemer's by two years and the American
courts decided in favour of Kelly on the question of American patent
rights. Nevertheless, it was the Bessemer/Mushet process that brought
about the first great expansion of the American steel industry, largely
owing to the mechanical superiority of Bessemer's converter.

in limited quantities in Sheffield by the crucible process developed by Benjamin Huntsman in 1746 and used primarily in the manufacture of knives and other edge tools. Instead, it was what we now term mild steel. As ductile as wrought iron, though not so readily firewelded, it was slightly superior in tensile strength to the best iron. Because it lacked the slag content of wrought iron, which gave the latter its fibrous quality, it was more subject to oxidation and this would prove a disadvantage in ship-building and in civil engineering applications such as bridge-building where the metal was exposed to the weather. But the Bessemer process promised that this 'steel' could be produced in uniform quality cheaply and in quantity. 'Promised' is here the operative word, however, for trouble lay ahead for Bessemer.

Although it may have seemed to Nasmyth that Bessemer's paper fell on deaf ears at Cheltenham, news of his discovery spread, causing a furore in the iron industry, and within weeks Bessemer had received £27,000 in licence fees from ironmasters anxious to use the new process. But, try as they would, these licensees produced metal worse than the poorest quality of wrought iron and completely useless. What was hailed as an outstanding invention was now denounced by *The Times* as 'a brilliant meteor that had flashed across the metallurgical horizon, dazzling a few enthusiasts, and then vanishing forever in total darkness'.

This fiasco was due to two troubles, the presence of excess oxygen in occluded form in the metal, a trouble which Nasmyth had earlier experienced, and the presence of phosphorus in the ores used which made the metal crack when worked under heat – or become 'hot short' in the ironmaster's graphic phrase. As luck (or ill-luck) would have it, Bessemer in his experiments at Baxter House had quite fortuitously used a Blaenavon pig iron made from one of the few British ores that was almost entirely free from phosphorus. Nevertheless, it was his lack of metallurgical experience that had led him to announce a victory when in fact the battle was only half-won.

The problem of how to get rid of the excess oxygen was solved by Robert Mushet, the son of a Scottish metallurgist and the owner of a small ironworks near Coleford in the Forest of Dean. In this remote works in the heart of the Forest, Mushet made a number of important contributions to the art of steel-making which were ill-rewarded, won him little renown in his lifetime, and have never been sufficiently widely recognized. When Mushet was presented with a sample of Bessemer metal made by one of the unsuccessful licensees, Thomas Brown of the Ebbw Vale Ironworks, he quickly diagnosed the trouble and prescribed its cure. Knowing from his own experiments that oxygen had a particular affinity for manganese, he suggested the addition of a proportion of manganese ore to the melt. This expedient proved completely successful. The surplus oxygen was attracted by the manganese and passed off in the slag as manganese oxide. Simple though this solution was – or perhaps because it was so simple – it was received with considerable scepticism, so Mushet decided to demonstrate its practicability himself. A small Bessemer steel plant was built for him by Fox, Henderson & Co., of Smethwick, the builders of the Crystal Palace, and installed at Mushet's Coleford Works. With this plant the first successful charge of Bessemer/Mushet steel was run in 1857 and the first ingot thus produced was successfully forged into a chopper by a young Free Miner of Dean Forest named George Hancox.*

* George Hancox died in 1922, aged eighty-six. His most prized possession was a book inscribed by his old master Mushet in these words: 'To Mr George Hancox, a first-rate intelligent British workman, an honour to the class to which he belongs. He forged the first commercially successful ingot of Bessemer steel ever made. . . .'

Nor was this all. A forty-seven-pound ingot of steel was cast at Coleford and sent to the Ebbw Vale Ironworks where it was successfully rolled into a short experimental length of double-headed rail. This was despatched to Derby where it was laid by the Midland Railway Company on a particularly busy crossover at the approaches to Derby station where the iron rails had to be renewed every six months, or occasionaly every three months. Ten years later, in 1867, Mushet inquired how many trains had passed over the rail daily and whether the Company would sell the rail back to him. To this the Midland Company replied that an average of 500 trains a day had passed over the rail and, though they were not prepared to sell it, he would have the first refusal should they ever decide to take it out.

With Mushet's vital contribution, Bessemer's process was completely successful, provided non-phosphoric ores were used; but although Bessemer reaped a fortune from his invention he never acknowledged Mushet's part in it. Much of Bessemer's wealth came from foreign licences, for in Britain the use of his process was limited by the fact that most British ores were phosphoric and good Bessemer steel could only be made from the non-phosphoric haematite ores of Monmouthshire, Cartmel and Cumberland. This situation lasted until 1879 when the problem was solved by Sidney Gilchrist Thomas, a Welshman who worked as a police court clerk in London, with the help of his cousin, P.C. Gilchrist, a chemist at Blaenavon Ironworks. Thomas's solution was to line the blast furnace with a form of limestone known as dolomite. Whereas the type of lining used heretofore was siliceous and acid and was therefore powerless to attract the phosphoric acid in the ore, the dolomite was alkaline and drew off the phosphorus to form basic slag. This became known as the *basic* process of steel-making as opposed to the older Bessemer *acid* process.

Long before the basic process of steel-making had appeared on the scene an alternative to the Bessemer method had been evolved. In 1857, William Siemens announced in a paper to the Institution of Mechanical Engineers a new type of regenerative furnace invented by his brother Freidrich. The brothers were Germans who had originally been employed by Fox, Henderson & Co. William later established himself as a consulting engineer in London, becoming a British subject and one of the most eminent engineers of his day. The Siemens furnace consisted of a heating chamber flanked by two hearths from which the hot gases passed into two regenerator chambers of honeycomb firebrick below where they heated the incoming air used in combustion. The air and gas flows could be reversed so that the chambers were alternately heated and cooled. In this way great economy was achieved, the furnace using only twenty to thirty per cent of the fuel consumed by one of conventional type.

The Siemens furnace was first used in 1861 for the making of glass by Messrs Chance Brothers of Smethwick who, it will be remembered, had supplied the glass for the Crystal Palace ten years before. In France two years later, however, Pierre and Emile Martin, father and son, discovered that they could make steel by melting pig iron with scrap steel and other suitable ingredients in a Siemens furance. This process took from six to fifteen hours as compared with thirty minutes for a Bessemer converter 'blow', but it had particular advantages. Whereas the converter could only accept molten metal from a blast furnace or a remelting cupola, the Siemens furnace could accept either molten or cold metal so that steel-making became a self-contained process. Moreover, the Siemens-Martin or open-hearth method of steel-making, as it came to be called, enabled the process to be far more accurately controlled. This last became an increasingly important consideration as methods grew more scientific and the grades of steel produced began to multiply by the use of various alloys.

William Siemens had difficulty in getting the process accepted in England so, like Robert Mushet, but with greater success, he set up a small demonstration plant known as the Sample Steel Works in a rented factory in Birmingham in 1866. Open-hearth steel-making on a commercial scale began in 1869 at the Swansea works of the Landore Siemens Steel Company, and by the end of the century three-quarters of Britain's annual steel output of nearly five million tons was produced by the open-hearth method.

As steel production rose, so the demand for wrought iron declined and the Black Country, having nearly exhausted its local supplies of iron ore, ceased to be the great iron-producing centre of Britain. Steel production tended to concentrate more and more in Sheffield and in a few large plants established wherever natural resources of low-grade ores could be conveniently exploited, such as at the Frodingham Ironworks in north Lincolnshire which came into blast in 1865.

No sooner were these new methods of economical steel-making adopted than the production of special alloy steels began. For it was found that by the addition of varying proportions of other metals – manganese, nickel, vanadium, tungsten, cobalt and chromium – to the furnace, the resulting steel could be endowed with different, but equally valuable properties. It could be made tougher and more resistant to wear; its tensile strength could be increased; it could be made harder or more heat-resistant.

As long ago as 1840, David Mushet, the father of Robert, had made ferro-manganese upon a small scale and had recognized its wear-resisting properties, but it was Sir Robert Hadfield who pioneered the large-scale production of manganese steel at Sheffield in 1887 and it was widely used for railway wheel tyres and those parts of railway pointwork that were subjected to the heaviest wear. Manganese steel was also non-magnetic, but although both Mushet and Hadfield recognized this property, it was only later that it proved of value in the electrical engineering industry.

Besides his contribution to the Bessemer process, Robert Mushet became a pioneer worker in the field of alloy steels. In 1862 he formed the Titanic Steel and Iron Company, and it is interesting to note that three members of the Pease family of Darlington, which had earlier promoted the Stockton & Darlington Railway and backed George Stephenson, were among the original subscribers. This emphasizes the extent of railway interest in new metallurgical developments. Under the new company's aegis, the Titanic Steel Works was established in Dean Forest. This was a small crucible steel-making plant where, in great secrecy, Robert Mushet carried out his alloy steel experiments. The name 'Titanic' was adopted, not from any illusions of grandeur, but because titanium was one of the alloys with which Mushet was experimenting at this time. So named by its German discoverer in 1794 from the Titans of Greek mythology because of its great strength, it is only in recent years that titanium has fully proved its worth, particularly in the aircraft industry.

Mushet's greatest success was a special self-hardening tool steel of immeasurably superior cutting power to the carbon steel used previously. Its superiority was first demonstrated in the machine shop of John Fowler & Co., of Leeds and 'R. Mushet's Special Steel', or 'R.M.S..' as it was stamped, soon became famous in machine shops on both sides of the Atlantic. Following his unfortunate experience with Bessemer, Mushet took the most extraordinary cloak and dagger precautions to keep his R.M.S. formula secret. The ingredients were always referred to by cyphers and were ordered through intermediaries so that their ultimate destination could not be traced. Even after the manufacture of R.M.S. had been transferred to Samuel Osborne & Co., of Sheffield, the mixing of the ingredients was still carried out in the seclusion of the Forest by Mushet himself and a few trusted men. These ingredients, mixed and partially processed, were then packed in barrels and consigned to Sheffield by varying roundabout routes and through futher intermediaries. Consequently Mushet's production methods remain unknown to this day, though subsequent analysis has proved that 'R.M.S.' contains proportions of manganese and tungsten with the addition of chromium from 1872 onwards.

After 1900, Mushet's 'R.M.S.' was superseded by other high-speed tool steels from America which were succeeded in turn by the tungsten-carbide tools originating in Germany. But Mushet's steel was the first in the field and brought about the biggest revolution in the engineer's machine shop that had occurred since the days of Maudslay and Whitworth. For machine tools everywhere had to be re-designed in order to take full advantage of the far greater cutting power of the new tool steel. The revolution was timely, for with the old carbon steel tools, components made from the new, tough alloy steels that were rapidly coming into use could never have been machined at all.

This story of the introduction of steel-making in Britain has been told in some detail because it caused changes of a most fundamental kind throughout the entire engineering world and thus indirectly affected the whole of manufacturing industry. In its far-raching effects it is comparable with such revolutionary innovations as coke smelting, the puddling process and the use of steam power. But of all these fundamental contributions to the Industrial Revolution to be made by Britain this was the last. Steel-making was the swan-song of the workshop of the world. It was rapidly taken up in Europe, notably by Germany and Belgium, while America, with her vast resources would, by

1900, out-strip Britain as the world's largest steel producer. Of the improvements in British steel-making technique introduced after 1900, the majority were of American origin. So, in considering the changes and developments in British engineering in the age of steel, it must ever be borne in mind that, unlike the pioneering age of iron, these were not necessarily novel but were paralleled in other countries, most notably in America.

In the structural use of steel, Britain made a late but spectacular use of steel. A committee was appointed by the Institution of Civil Engineers to investigate its properties in 1868, but five years later the engineer W.H. Barlow was complaining to the British Association at Bradford that:

> We know that it [steel] is used for structural purposes in other countries, as, for example, in the Illinois and St Louis bridges in America, yet in this country, where *modern steel* has originated and has been brought to its present state of perfection, we are obstructed by some deficiency in our own arrangements and by the absence of suitable regulations from the Board of Trade from making use of it in engineering works.

However, bureaucracy belatedly sanctioned its use in time* for steel to play a notable part in that most dramatic story of engineering tragedy and triumph, the Tay Bridge disaster and the successful bridging of the Firth of Forth which followed it.

This story had its origins in the keen competition for traffic between the railway companies whose lines comprised the east-and west-coast routes to Scotland.

---

* According to Barlow, steel girders were first used in Britain on the Chester & Holyhead line in 1883, presumably in the renewal of underline bridges.

1. Brunel at the launch of the *Great Eastern*. Brunel, Stephenson and Locke (following pages), the great triumvirate of railway-builders.

2.   Robert Stephenson

3.   Joseph Locke

4. Robert Stephenson's High-Level Bridge over the Tyne at Newcastle, completed 1849. The bridge still carries rail and road traffic on its two decks. From a lithograph by George Hawkins after J.W. Carmichael.

5. The Dee Bridge disaster of 1847, from an *Illustrated London News* picture. The bridge was designed by Robert Stephenson and the collapse under a train of one of its trussed cast-iron girders with the loss of five lives caused a furore in engineering circles. As a result, the use of cast iron in railway structures fell into disfavour.

6.   The great rectangular tubes for Robert Stephenson's bridge over the straits of Menai under construction on staging built along the Caernarvon shore. The uncompleted piers of the bridge appear in the background, 1848. A hand-coloured lithograph from a drawing by S. Russell.

7.   The 'Great Cylinder' which was used to build the central pier of Brunel's Royal Albert Bridge at Saltash. The cylinder has been towed out into the centre of the Tamar and is about to be up-ended and sunk. From a contemporary lithograph.

8. The ceremonial opening of Brunel's Royal Albert Bridge at Saltash by the Prince Consort, 1859. From the oil painting by Thomas Valentine Robins.

9.   A group of engineers, contractors' staff and foremen pose before the almost completed Severn Bridge, August 1879. The resident engineer, G.W. Keeling, is the central figure leaning upon his elbows.

10. Midland grandiloquence, St Pancras. This interesting watercolour by an unknown artist was evidently painted before the design of the buildings had been agreed upon. It shows engineer W.H. Barlow's great single-span roof substantially as built, except that its arch never bore the station names as shown. The design of the buildings was the subject of a competition and those shown here may represent the proposals of one of Sir Giles Gilbert Scott's unsuccessful rivals.

11. Victorian temple of steam: Whitacre Pumping Station, Staffordshire. The engines are inverted compound beam engines by James Watt & Co. They commenced working in 1885, ceased in 1937 and were dismantled in 1950. Notice the corbels in the form of gilded eagles which supported the inspection galleries surrounding the upper cylinder covers.

12. The first Atlantic steamship: Brunel's P.S. *Great Western* is towed out of Patterson's Dock at Bristol preparatory to voyaging under sail to the Thames to take on her Maudslay engines, July 1837. From a lithograph by J. Walter.

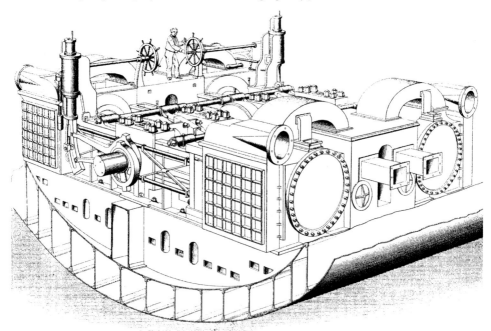

13. Drawing from Bourne's *Treatise on the Steam Engine* (1862) showing the screw engines of the Great Eastern of 1859. Built by James Watt & Co., this was then the most powerful unit in existence. Two pairs of seven-foot-diameter cylinders were horizontally opposed and, using steam at twenty-five pounds per square inch, together indicated nearly 5,000 h.p. Note the vertical steam cylinders used to operate the Stephenson link reversing gear.

14. Premature Leviathan: Brunel's *Great Eastern* of 32,000 tons displacement, 1859. From a lithograph by T.G. Dutton.

15. Brunel's great ship comes into her own as a cable layer: 'Sunrise on the deck of the *Great Eastern* on 2nd September 1866 after retrieving the lost Atlantic cable', from the oil painting by an unknown artist in the possession of the Institution of Civil Engineers. The picture shows the aft deck with its cable-laying machinery.

16. Crystal Palace

17. The 'so-called' 'aeronautic view' of the Great Exhibition building in Hyde Park as seen from the north-west. The best contemporary illustration of Paxton's 'Crystal Palace'.

18. The scene on May Day, 1851, as the Queen declared the Great Exhibition open. The view shows Osler's crystal fountain at the junction of the naves and transepts.

19. Transferring the girders from the old to the new Tay Bridge, 1885–6. One of the piers of the old bridge designed by Sir Thomas Bouch can be seen on the extreme right and should be contrasted with those of W.H. Barlow's new bridge.

20. The Triumph of Steel: The Forth Railway Bridge of 1890. From the original oil painting commissioned in memory of the designer of the bridge, Sir Benjamin Baker, and presented to the Institution of Civil Engineers.

21. 2,000 h.p. unleashed: Charles Parson's experimental turbine ship *Turbinia* travelling at a speed of over thirty knots when on her trials. *Turbinia* is now preserved in the Newcastle Museum of Science & Engineering.

22. Two great engineers of the late Victorian age (left and below): the Hon. Charles A. Parsons.

23. Sebastian Ziani de Ferranti.

The long, direct and level miles between London and York gave the east-coast companies a decisive advantage as far as the Scottish border, for the west-coast route was longer and included Locke's killing-climb over Shap Fell. North of the border, however, it was quite otherwise, for the east-coast passenger found his journey interrupted by two slow and sick-making ferry voyages over the Firths of Forth and Tay. So the east-coast companies resolved to bridge the Forth and Tay. The North British Railway undertook the latter, while a consortium was formed to bridge the Forth, and Thomas Bouch was appointed engineer. Bouch prepared plans for both bridges and a contract for the Tay Bridge was let in 1871. Consisting of wrought-iron lattice girder trusses supported on cast-iron columns with masonry bases, the bridge was completed and opened in 1878. In the England of the 1870s, railway construction no longer appeared the heroic adventure that it had been in the pioneering days of Stephenson and Brunel, but this great bridge, by far the longest in the world, was really something special. It was opened by the Queen who knighted Thomas Bouch on the spot. Little over a year later a whole section of Bouch's bridge was a tangled mess of girders in the bed of the Firth. It had blown down in a December gale carrying an entire train and its complement with it.

A complacent Victorian public was stunned by the news of this disaster. It was regarded as a serious blow to national prestige. A public inquiry was held which reported:

> We find that the bridge was badly designed, badly constructed and badly maintained and that its downfall was due to inherent defects in the structure which must sooner or later have brought it down. For these defects both in design, construction and maintenance Sir Thomas Bouch is in our opinion mainly to blame.

Poor Bouch was a ruined man. That there were serious faults in design and construction is undeniable, but Bouch's fatal error was that, in preparing his design, instead of trusting to his own judgement or carrying out practical tests on the spot, he had consulted the Astronomer Royal, Sir George Airey, as to what wind pressure he might expect on the structure. That worthy had replied: 'The greatest wind pressure to which a plane surface like that of the bridge will be subjected in its whole extent is 10 lbs per square foot.' In view of Airey's earlier dogmatic pronouncements, Bouch should have known better than to trust such a statement.

Needless to say, preliminary works for a bridge of Bouch's design over the Forth were immediately stopped, but matters were not allowed to rest very long. British civil engineers were eager to retrieve their profession's damaged reputation. Work on rebuilding the Tay Bridge to the designs of W.H. Barlow commenced in 1882. The girders of Bouch's bridge were traversed onto new piers. New girder spans of wrought iron with steel decks were floated out into the firth to replace those which had fallen and the bridge was completed in 1887.

Meanwhile Sir John Fowler and Benjamin Baker, consulting civil engineers, were asked to submit a design for a bridge over the Forth and to superintend its construction. Fowler was responsible for the masonry approach viaducts and Baker for the main structure. The site settled upon for the crossing was at Queensferry where the rocky islet of Inchgarvie offered a base for a mid-water pier. Indeed the task confronting Baker was precisely the same as that which Robert Stephenson had earlier faced in the crossing of the Menai Straits except that it was magnified fourfold. For whereas two main spans of 460 feet each had sufficed for the Menai crossing, here at the Forth the two spans required were each 1,750 feet.

Unprecedented problems call for novel and daring solutions. Just as Stephenson had resolved to employ wrought iron upon a scale never seen before, so now Baker determined to use steel in a manner equally grand. In the same year that Henry Bessemer had announced his invention, Benjamin Baker had entered the Neath Abbey Ironworks as a young apprentice, so he had practical working knowledge of iron- and steel-making.

Baker's first act was to set up a wind gauge on Inchgarvie where a maximum velocity representing a pressure of thirty-four pounds per square foot was recorded. So much for Sir George Airey's prediction. Nevertheless, Baker designed his bridge to withstand a wind pressure of fifty-six pounds per square foot which amounted to a lateral load of 2,000 tons on each of the main spans.

The largest size of wrought-iron plate available to Stephenson in building his Britannia Bridge had measured twelve by two feet and all wrought-iron beams used had to be riveted up from plate and angle sections. Now, however, the large size of steel ingots, coupled with great improvements in the design of rolling mills, enabled the steel-maker to supply Baker with much large plates and a great variety of rolled steel sections including I-section girders of large size. The steel for the Forth Bridge was produced by the Siemens open-hearth process and the steel-makers of Scotland supplied plates as large as thirty by five feet. But the really crucial factor was that this steel had a tensile strength of thirty-three tons per square inch as compared with twenty-two tons for the wrought iron used by Stephenson. As Sir John Fowler remarked after the Forth Bridge had been completed: 'With iron it would have been twice the weight (if practicable at all), in consequence of the less strength of the material, and more than twice the cost, and therefore impracticable. . . .'

At Queensferry, the contractors, Messrs Tancred Arrol, installed special plate rolling, planing and multiple drilling machines by the aid of which the steel plates were shaped into the great tubes that formed the main members of the cantilevers and of the three towers that rose 343 feet into the air from the piers near the Fife and Queensferry shores and on the rock of Inchgarvie. From these towers the cantilevers outspread their huge arms until, at their extreme reach of 680 feet from the piers, they upheld between them, high above the waters of the firth, the two suspended spans. After eight years' work, the Forth Bridge was completed and British civil engineering skill was vindicated. The bridge was opened by the Prince of Wales on 4 March 1890 and Benjamin Baker received a Knighthood. '. . . . I do not believe in astronomy being a safe guide for practical engineering', remarked Sir John Fowler dryly, as he remembered the fate of poor Thomas Bouch.

In 1890 the Forth was the largest bridge in the world, and as great a pioneer achievement as the Britannia bridge had been forty years before. Today both are historic monuments of the first importance, marking the dawn of the ages of steel and iron. But times had changed in forty years. Unlike the pioneer generation of engineers whose works the artist had been delighted to honour, Sir Benjamin Baker was forced to defend his bridge against hostile criticism. William Morris condemned it as 'the supremest specimen of all ugliness', and whereas a generation earlier Paxton's Crystal Palace had won nothing but praise, now Morris declared roundly that 'There never would be an architecture of iron, every improvement in machinery being uglier and uglier'.

> Probaly [Baker replied], Mr Morris would judge the beauty of a design from the same standpoint, whether it was for a bridge a mile long or for a silver chimney ornament. It is impossible for anyone to pronounce authoritatively on the beauty of an object without

knowing its functions. The marble columns of the Parthenon are beautiful where they stand, but if we took one and bored a hole through its axix and used it as a funnel for an Atlantic liner it would, to my mind, cease to be beautiful, but of course Mr Morris might think differently.

These were the opening shots in the battle between the engineer and those concerned to defend the beauties of the English landscape which has been going on from that day to this. Sir Benjamin Baker, like William Morris, had a deep feeling for the past. He revered the memory of his great predecessors, particularly Thomas Telford, several of whose works he sympathetically restored. Yet in this exchange we see two great men, each famous in his own sphere, diametrically opposed and each incapable of seeing the other's point of view. Both had right on their side. Baker the engineer was right in his insistence upon structural honesty. Morris the artist was right in pointing out that the works of the engineer were becoming uglier and uglier. He found the cantilevers of the Forth Bridge as ugly as we do the giant pylons of today and for the same reason: because they revealed to him how rapidly man's growing mastery over matter was alienating him from his natural environment.

It would be interesting to know what these two formidable antagonists thought of the Tower Bridge, which was completed in 1894, two years before Morris's death. For this huge hydraulic machine masquerading as a medieval drawbridge is the supreme example of the work of an engineer, Sir John Wolfe-Barry, 'civilized' by an architect, Horace Jones.

Critics notwithstanding, with his Forth Bridge Baker dramatically demonstrated to the world at large the civil engineering potential of steel. In every other branch of the engineering profession steel made possible equally important changes and developments even if the results, or the reason for them, may not have been evident to the lay public. To boiler-maker and ship-builder steel spelled stronger and lighter boilers and hulls with fewer riveted joints. Boiler pressures rose as a result, and ships increased rapidly in size.

John Ramsbottom, Locomotive Superintendent of the London & North Western Railway and inventor of the water-pick-up apparatus, began making steel locomotive boilers at Crewe in 1873 and by the 1890s even the most conservative boiler shops had turned to steel. Earlier, in 1866, Ramsbottom began producing steel rails at Crewe using a reversible rolling mill of his own design. Although the steel rail was introduced for its hard-wearing qualities, it also brought about as great an improvement in travelling comfort as modern welded track. For whereas the earlier wrought-iron rails had been rolled in twenty-one feet lengths, the new Crewe mill could produce steel rails sixty feet long so that the number of rail joints could be reduced proportionately. The L.&N.W.R. was unique in rollings its own rails, but other railway companies soon followed suit, purchasing rails from steel-makers who specialized in rail rolling as did those of Ebbw Vale and Dowlais in South Wales. Robert Mushet's short experimental rail at Derby station heralded a rapid revolution that soon made the nineteenth-century expression 'the iron road' a misnomer.

The use of steel in place of iron for such highly stressed parts of locomotives as coupling rods, connecting rods, crank-pins and axles, together with the introduction of wear-resistant cast-steel wheel tyres made possible a revolution in locomotive design. The fact that the use of steel boilers brought about a general increase in working steam pressures also contributed to this.

Uneven wear of the iron tyres in locomotives had hitherto set up abnormal stresses in coupling rods, crank-pins and axles which not infrequently led to failures, sometimes disastrous. Hence the preference of locomotive engineers for 'singles' (locomotives with a single pair of driving wheels) for express passenger service, coupled engines being largely confined to slow-speed goods services. Despite the introduction of steel, this preference persisted among the more conservative locomotive engineers until almost the end of the nineteenth century and the single express engines of Patrick Stirling of the Great Northern, Samuel Johnson of the Midland and William Dean of the Great Western were, in the opinion of many, among the most beautiful locomotives ever built. Such men justified their preference by extolling the free-running qualities of the single, that dry Scotsman Patrick Stirling contempt-uously dismissing the performance of coupled engines as 'like a laddie running' wi' his breeks doon'.

But insistent demands for greater space and comfort by a new generation of railway travellers who had never known the rigours of stage-coach travel enforced a change. More roomy eight-wheeled, and even twelve-wheeled, coaches to which the amenities of side corridors, dining- and sleeping-cars were soon added, greatly increased train weights and sounded the death knell of the supremely elegant single express locomotive. Stirling, its most stubborn advocate, built his last singles at Doncaster in 1894–5, but it is significant that in order to make them cope with heavier trains he increased the adhesive weight on the single driving axle to an excessive extent to which two subse-quent derailments were partially attributed.

By this date most locomotive engineers had accepted the fact that for the heaviest express service the use of four coupled wheels in association with a leading bogie (4–4–0) was essential. The first two engines of this type* in Britain

---

* One of these two pioneer locomotives hauled the ill-fated train that was lost in the Tay Bridge disaster in December 1879. She was subsequently recovered, little worse for wear, and worked until 1919.

had been designed by Thomas Wheatley for the North British Railway in 1871 and the type was developed in 1876 by his successor Dugald Drummond who, with his brother Peter Drummond, was chiefly responsible for popularizing it on Britain's railways in subsequent decades. Stirling's successor at Doncaster, Henry Ivatt, and John Aspinall of Lancashire & Yorkshire Railway were responsible for making the next step forward, by introducing designs of 'Atlantic' type (4–4–2) express locomotives in 1898 and 1899 respectively. The extra pair of wheels enabled a much larger boiler to be carried. The performance of such engines proved that, in an age of steel, the belief that no coupled engine could match the single for sustained high-speed running combined with reliability was a fallacy.

Meanwhile, on the Highland Railway in Scotland, David Jones had introduced in 1894 a mixed traffic locomotive with leading bogie and six-coupled wheels (4–6–0). This foreshadowed the many designs of six-coupled express locomotives which would appear on Britain's railways in the first decade of the twentieth century in order to cope with ever-increasing train weights.

By making possible higher steam pressures, the efficiency of the steam engine in all its applications was greatly increased in the steel age. But no matter how high its pressure may be, steam if delivered direct from the boiler to the engine contains water vapour and is said to be in a wet or 'saturated' state. When such steam comes in contact with the cool surface of the cylinder, the water vapour condenses, causing a loss of economy. This bar to higher efficiency had long been recognized and so had the obvious remedy. This was to 'superheat' the steam in its passage between boiler and engine, but the realization of this seemingly simple cure had presented insuperable technical difficulties. Steam pipes of wrought iron or copper would not for long withstand the heat to which they were subjected. Also, increased working temperatures in the engine cylinder created serious lubrication problems and burned out the cotton or hemp packings with which piston rod and valve spindle glands were kept steam-tight. Heat-resistant pipes of alloy steel made superheating possible, while mineral oil, fed into the cylinder by displacement lubricators, replaced the old tallow pot lubricators. The introduction of spun asbestos in place of cotton from 1879 onwards and the subsequent development of metallic packings solved the gland problem. Consequently the successful introduction of superheating began in 1890, although it was not adopted on locomotives until the following decade. With the use of superheated steam at high pressure, the reciprocating steam engine was given a new lease of life and so was able to hold its dominating position to the end of the century and beyond. The first challenge to its supremacy came, not from the internal combustion engine, but from the steam turbine.

The principle of the steam turbine is similar to that of the water wheel, or its refinement the water turbine, and the origin of the idea can be traced back into prehistory. To harness either the reactive or impulsive force of a jet of steam in a similar manner seemed a much simpler way of obtaining rotary motion than that of the cylinder and piston which needed rods and cranks to convert linear into rotative motion. Numerous pioneers, including James Watt and Richard Trevithick, produced designs for so-called 'whirling engines', 'rotary engine' or 'steam wheels', but without success. The snag was this: to operate efficiently, the buckets or floats on a water-wheel have to revolve at half the speed of the water impinging upon them. Precisely the same law applies to the steam turbine; otherwise the pressure and expansive power of the steam cannot be utilized fully and the turbine will be woefully inefficient by comparison with an orthodox reciprocating steam engine. In practical terms this meant that, to be successful, the rotor of a steam turbine must revolve at an astronomical speed quite beyond the capacity of early nineteenth-century engineering to achieve. At a time when the speed of the reciprocating steam engine was measured in hundreds of revolutions per minute, the turbine would require to run at as many thousands. Not until the age of steel were there materials of sufficient strength and durability or methods of forming and machining those materials with the necessary accuracy to make the steam turbine a practical proposition.

The pioneer of the steam turbine was the Hon. Charles Parsons, sixth and youngest son of the Earl of Rosse, an unusual family background for a practical engineer at this time. He commenced his engineering training at the Elswick Works of Sir William Armstrong, and it was while he was still an apprentice there in 1880 that he first began to consider the problem of the steam turbine. His starting point was the simple water turbine, though he soon realized that the parallel between a rotor powered by an incompressible fluid like water and one driven by the kinetic energy of steam was not exact and that, however fast the rotor was made to revolve, it could not absorb more than a fraction of that energy. Still pursuing the analogy of the water turbine, however, Parsons argued that if a number of such turbines were arranged in series on a common shaft and the steam passed from one to another successively, the analogy would become much more exact. Each rotor would then be required to absorb only a small proportion of the energy of the steam and therefore rotational speed, though still high, would be kept within practicable bounds. Just as the earlier designers of reciprocating steam engines had improved their efficiency by expanding the steam in successive cylinders, so Parsons now proposed to expand it through successive turbine rotors. The idea held the promise of great efficiency, for whereas each cylinder added to the reciprocating engine spelled greatly increased complication and frictional losses, the multiplication of turbine rotors on a single shaft presented no such disadvantage. Used in association with a condenser, it would, in theory, be possible for such a multiple stage turbine to harness the whole of the pressure drop of the steam, from high pressure at the inlet to absolute zero at the outlet. Such a result the reciprocating engine could never hope to achieve. Its low-pressure cylinder could never be made large enough, while even if it were, it would be practically impossible for the exhaust steam to pass sufficiently quickly into the condenser, no matter how high the vacuum.

Parsons patented his steam turbine in April 1884. No patent of comparable importance had been taken out in the field of prime movers since Watt had patented his separate condenser in 1769. In both cases, though the idea was relatively simple, its realization in practical form was not and involved the solution of mechanical engineering problems which stretched contemporary techniques to the utmost. Of the two men, Parsons had the advantage of a better education, but both had the same scientific habits of thought combined with an intuitive genius for arriving at the correct solution of a problem. But Parsons had the added advantage of practical workshop training. In the words of his lifelong friend Dr Gerald Stoney, 'Not only was he endowed with a thoroughly practical type of mind, but he had also great mechanical aptitude. He was perfectly familiar with all workshop processes and had unusual skill with his hands.' This enabled Parsons to surmount the formidable technical problems that barred the way to the production of a successful steam turbine, whereas Watt had to rely on the practical help of others in translating his ideas into practice and was sometimes driven to despair by the difficulties involved.

The long-term social effect of the steam turbine has been almost as great as that of Watt's steam engine. Today, when the layman is apt to regard steam as an obsolete source of power belonging to the nineteenth century, an overwhelming proportion of the electrical power we use is generated by steam turbines operating as steam pressures and temperatures inconceivable to James Watt and which could seem astronomical even to Charles Parsons. Why then does the name of Parsons not enjoy the same popular fame as that of Watt? The answer is that in the 1880s British engineers did not stand upon the same peak of lofty eminence which they had occupied a hundred years before. Whereas Britain alone could bring Watt's steam engine to a successful birth, now other countries were as capable as Britian of acting as midwives to the steam turbine. Hence Parson's pioneer work was closely followed by that of Gustav de Laval* in Sweden, who produced his first design in 1887, and by the turbines of Rateau (1896) and Ljungstrom (1910) in Europe and Curtis (1897) in America. All these inventors employed different mechanical means to achieve the same end, but because these means are not readily distinguishable by the layman their effect has been to confer a certain anonymity on the invention of the steam turbine. It is but a short step from this to the situation which obtains today where invention has become to an increasing extent the province of teams of anonymous specialists.

It will be recalled that when the screw propeller was first introduced, driving it presented a problem to marine steam engineers owing to the extremely low rotational speed of their engines. They were forced to employ speed-up drives

---

* It is only fair to say, however, that de Laval's steam turbine in its original form was far less highly developed than was Parson's. With its single-vaned rotor impelled by multiple variable orifice steam jets, it was much nearer to the water turbine prototype and was successful only in small power applications.

to the propeller shaft. Now, however, the steam turbine presented precisely the opposite problem, for the first turbine to be built by Parsons revolved at 18,000 r.p.m. Parson's intention from the beginning was that his turbine should be used for generating electricity, directly coupled to the generator. The speed demanded by the generators then in use was 1,200 r.p.m. and it speaks volumes for the versatility of Parsons's genius that he was able to design a new dynamo suitable for his turbine, patenting it in the same year as the latter.

Not content with this, in 1896 Parsons built the first vessel in the world to be propelled by turbines, the small experimental *Turbinia*. Three propeller shafts, each carrying small triple propellers, were driven by high-, intermediate- and low-pressure turbines respectively, which together developed 2,000 h.p. A fourth turbine was coupled to a central shaft for reversing. The *Turbinia* demonstrated yet another advantage of the turbine over the reciprocating engine: its small size in relation to the power produced and the fact that in running it was vibrationless, requiring no massive foundations. Never before had it been possible to pack so great a concentration of horse-power into one small hull. With malicious glee, Parsons demonstrated this fact by turning his *Turbinia* loose among the ships at the 1897 Jubilee Naval Review at Spithead like some sheepdog in a field of sheep. To the consternation of the Navy's top brass, the little *Turbinia* tore between the assembled ranks of flag-be-decked ships. The fleet's fastest destroyers were ordered to intercept the intruder, but in vain. As she was capable of the hitherto unheard-of speed of 34½ knots, *Turbinia* showed them a clean pair of heels.

This historic little ship now occupies a place of honour in the Museum of Science & Engineering at Newcastle. Despite her convincing demonstration at Spithead, the commercial application of marine turbines did not follow until the very end of our period when the appropriately named turbine steamer *King Edward* began plying on the Clyde in June, 1901, where she was joined by the *Queen Alexandra* in the following year. The first ocean liners to be fitted with Parsons turbines were the *Victoria* of the Allan line (1904) and the Cunarders *Carmania* (1905), *Lusitania* and *Mauretania* (1907), the latter being of 70,000 h.p. All these ships were fitted with direct drive from the turbines to multiple propellers, as on the *Turbinia*. But although this system was admissible on passenger ships where speed of turn-around was essential to their economic operation, it was unsuitable for the general run of merchant shipping where speed was less important than operating economy. Parsons therefore introduced a geared turbine in which twin turbine units were linked by reduction gears to a single propeller shaft. His first small experimental set of geared turbines was built in 1897 and one of the first large ships to be fitted with such a power unit was the SS *Transylvania*.

That the speed of the steam turbine could be reduced in this way to a shaft speed of no more than 80–200 r.p.m. with complete reliability and without substantial loss in mechanical efficiency would have been unthinkable thirty, or even twenty years before. It was made possible in the age of steel, not only by new materials but by advances in the engineering machine shop where gears having mathematically correct tooth forms could be cut on new machines with a degree of accuracy unattainable before.

The steam turbine illustrates a general tendency in mechanical engineering for massive slow-speed power units and methods of power transmission to be replaced by lighter, smaller and more efficient units and systems which develop or transmit equal power but at higher revolutions. This process had been going on since the beginning of the Industrial Revolution. The introduction of steam power was itself an acceleration. Writing of the improvements made by Sir William Fairbairn to cotton-mill drives in Manchester in 1818, Samuel Smiles had this to say:

> In the course of a few years an entire revolution was effected in the gearing. Ponderous masses of timber and cast iron, with their enormous bearings and couplings, gave place to slender rods of wrought iron and light frames or hooks by which they were suspended. In like manner, lighter yet stronger wheels and pulleys were introduced, the whole arrangements were improved, and, the workmanship being greatly more accurate, friction was avoided, while the speed was increased from about 40 to upwards of 300 revolutions a minute.

The replacement of massive pumping engines by small electric submersible pumps, or the supercession of the reciprocating internal combustion engine by the gas turbine in the air and in a growing number of other applications, provide examples of the same process at work today.

But if a graph of this continuing trend could be plotted its most significant rise would be seen to follow the advent of steel and its alloys. Alloy steels made possible better and more accurate gears, ball and roller bearings and lighter, more highly stressed machine components of every kind. But although Britain led the world into the steel age, no single country can claim credit for the engineering developments to which it so swiftly led. In the design of improved machine tools, for example, the United States seized the initiative and played an increasingly dominant role with German and Swiss engineers rapidly overhauling that commanding lead in Europe which, thanks to Maudslay and Whitworth, Britain had once held. The Industrial Revolution had become a world-wide movement.

EIGHT

New Lamps for Old

Mere mention of the name of Victoria conjures up a vision of an urban street scene of hansom cabs jingling past through a yellow, smoke-laden fog in which the pale glare of incandescent gas lamps glimmers fitfully. In fact, such a scene owes more to the imagination than to reality for it was conjured up by Sir Arthur Conan Doyle as an appropriate setting for the adventures of Mr Sherlock Holmes of Baker Street. It was only true to life over a very short period of time; as a general impression of the Victorian age it is misleading.

The incandescent gas mantle, for example, as invented by the Austrian, C.A. von Welsbach, was not adopted by British gas companies until the 1880s in an attempt to counter the competition from electric lighting which was already becoming serious. It was the Welsbach mantle that gave the light whose quality we now think of as essentially Victorian, whereas for the greater part of the nineteenth century, town houses and streets were lit by the dimmer and yellower glow of batswing or fishtail burners which smelt abominably and blackened ceilings. The fishtail burner, incidentally, was the invention of James Neilson who was foreman of Glasgow gasworks before he patented his hot blast stove. Poor though the light from these gas burners was, it was a vast improvement on the crude whale oil lamps, the candles and the rushlights of an earlier era and so contributed not a little to the spread of literacy. Before the age of gas light, most of the working population, even if they were literate, retired to bed in winter when their long day's work was done.

The introduction of gas lighting in England dates from the beginning of the nineteenth century and, like that of steam power, it owes much to that famous firm of Boulton & Watt of Birmingham. Experiments with coal gas for lighting had earlier been made in England, Belgium and, more particularly, by Philippe Lebon in France, but for the first plant to produce gas upon a commercial scale the credit goes to William Murdock, Boulton & Watt's chief engine erector in Cornwall. There he built a small experimental plant with which he successfully lighted his house in Redruth. When he was recalled to Birmingham in 1798 to become manager of the Soho Foundry, he installed there a gas-making plant large enough to light the works. Moreover, to celebrate the Peace of Amiens in March 1802, the façade of the original Soho Manufactory was floodlit by gas flares supplied from this plant, a spectacle which excited much wonder and admiration.

Gas light first appealed to the owners of textile mills as offering a form of lighting safer than candles or oil lamps, so reducing fire insurance premiums. Because it also provided a better and more reliable light, it enabled mills to work by night as reliably as by day, thus greatly increasing the mill-worker's burden of toil.

By 1804 Boulton & Watt were in a position to offer to undertake gas-lighting installations. Their first large order came in 1806 from Messrs Phillips & Lee, cotton spinners of Salford. These owners were evidently particularly conscious of the fire risk, for their large mill at Salford had been built for them by Boulton & Watt in 1801 on 'fire-proof' lines using cast-iron columns and beams in place of wood.* It was in this building that Boulton & Watt installed a battery of six retorts and 900 gas lights.

Boulton & Watt had at this time a commanding lead in this new field; the ingenius Murdock was constantly improving the apparatus and the Soho Foundry was admirably equipped to manufacture it. Yet for some unaccountable reason they failed to exploit this situation, dropping out of the new industry on the very eve of its expansion. This may have been connected with the death of Matthew Boulton in 1809, for there can be no doubt that with the passing of this most remarkable and dynamic man a great deal of the steam went out of his business.

The initiative now passed to Samuel Clegg who had joined Boulton & Watt as an apprentice in 1802 and after his departure in 1805 began to install factory gas plants on his own, the first being in Halifax. He later teamed up with a German immigrant named F.A. Winzer (later changed to Winsor) who had conceived the idea of a public gas supply from a central gererating station. His first effort, which was not very successful, was to install lights in Pall Mall. The London Gas Light & Coke Company was founded by Winsor in 1812 and, with Clegg's engineering aid, the parish of St Margaret's, Westminster, was lit by gas from works in Cannon Row two years later. By 1823, London already had 122 miles of gas mains, rising to 600 miles in 1834, and by the time Victoria came to the throne gas lighting was securely established, not only in London but in the larger provincial towns. It had also spread, or would soon do so, to the major cities of the world. Yet Britain led,

---

* William Fairbairn stated that this was the first building of its kind to be erected. This may have been true for Lancashire, but he evidently had not heard of Bage's earlier building at Shrewsbury. (See S.B. Hamilton, 'The Use of Cast Iron in Building', *Newcomen Society Transactions*, Vol. XXL, 1940–41.)

in this field as in so many others, because nowhere else was there such a concentration of population or such ample supplies of cheap coal.

The use of gas as a convenient source of heat was known quite early in the century and the gas cooker had appeared by the time of the Great Exhibition in 1851. But the emphasis initially was all on lighting, and heating applications were comparatively rare. It was not until towards the end of the century when electricity began to challenge gas as an illuminant that the gas companies began seriously – and successfully – to exploit the heating value of gas and gas cookers, radiant fires and geysers were widely introduced. It was at this time too, as we shall see, that the gas main also became a source of power transmission through the medium of the gas engine, a development that would have momentous consequences.

The introduction of electricity, first as an aid to communication, then for lighting and finally as a means of transmitting power, marked a significant change in the relationship between science and engineering. Hitherto, with the possible exception of refrigeration, engineers had been primarily responsible for all the major advances that have been described in this book. Although they may sometimes have made use of scientific methods in the course of their experiments, they were essentially practical men who achieved their results by the combination of trial and error, practical experience and inspired intuition. In the headlong progress that the engineer initiated in this way the scientist played a very small part. Plodding rather breathlessly in the engineer's wake, his usual role was to expound the theory of a machine or a process after the engineer had invented it. Now, however, the roles were decisively reversed. The vast electrical industry and the new branch of the engineering profession which served it could never have come into being but for the theories and the laboratory experiments of scientists working not only

167

in Britain, but in Germany, France, Italy and Denmark. From primitive seventeenth-century experiments with static current generated by friction, these men of many countries, each advancing the work of his predecessors, gradually added to the sum of knowledge until, by the middle decades of the nineteenth century, the stage was reached when that knowledge could be applied by the engineer and laboratory apparatus translated into practical machines. The new discovery required a new language of its own and thus the names of some of these pioneer scientists, Alessandro Volta and A.M. Ampere of Italy and George Simon Ohm of Germany, together, somewhat irrelevantly, with that of James Watt, became in the new electrical age part of common speech. They were adopted at an international conference held in Paris in 1881. Electrical engineering thus evolved by a process of international scientific collaboration which was then unique but has since become the normal process of technical advance. The only difference is that whereas modern communications have made such collaboration a rapid and organized process, in this pioneer example it was unorganized, fortuitous and therefore slower.

Electric current was first produced by electro-chemical means in the shape of the Voltaic pile (1800) and the Daniell (1836) and Leclanché (1866) cells. Batteries capable of storing electricity fed into them were a French invention and did not make their appearance until the late 1870s, but meanwhile the development of the so-called 'dry-battery' had made electric telegraphy possible.

The pioneers of electric telegraphy in England were Professor C. Wheatstone and W.F. Cooke. They developed the apparatus evolved by von Soemmering and Schilling in Germany and patented it in England in June 1837. The partners correctly foresaw that the need to control a growing railway traffic in the interests of safety offered the most promising application for their invention. They therefore obtained permission from the London & Birmingham Railway Company to install their apparatus experimentally between the Euston terminus and the engine house at Camden Town which was then used to draw the trains by rope haulage up the steep gradient out of the station.

Although this first telegraph worked satisfactorily, the railway company inexplicably failed to adopt it, clinging stubbornly to the method of signalling by pneumatic warning whilstle whose eerie wailings continued to be a disturbing feature of Euston station until the rope haulage system was finally abandoned. But if one railway company remained blind to the advantages of the telegraph, another did not. I.K. Burnel recommended to the Board of the Great Western Railway that it be installed between Paddington and West Drayton, a distance of thirteen miles. It was working as far as Hanwell by April 1839, and was extended to Slough by 1843. It achieved much celebrity in August 1844 by conveying to London the tidings of the birth of the Queen's second son, the Duke of Edinburgh, at Windsor.

It won ever greater fame the following year by making possible the prompt capture and conviction of a murderer. John Tawell had travelled down to Slough with a phial of cyanide in his pocket with the intention of poisoning his mistress. This he achieved, but his wretched victim gave such a piercing scream as she expired that she alerted her neighbours. Seen boarding a London-bound evening train at Slough station, a description of Tawell was immediately telegraphed to Paddington and led to his arrest and convinction.

The early telegraph instruments were 'speaking' instruments, that is to say their needles indicated the letters of the alphabet, enabling words to be laboriously spelled out. Certain of the more rarely used letters were omitted, and the fact that there was no 'Q' presented a problem when Tawell's description, 'a man in the garb of a Quaker', had to be transmitted. The resourceful clerk at Slough solved it by sending the word 'KWAKER'. On the instrument he used, twenty letters were arranged on a diamong-shaped board in such a way that the five needles by their inclination could indicate any of the letters at the point of intersection of the convergent lines between any two needles. The five wires required were carried in a tube mounted at ground level beside the lines.

Before the code which bears his name was invented by the American, Samuel Morse, simpler double and single needle 'speaking instruments' had been introduced with a proportionate reduction in the number of wires. At the same time the wires were carried on poles instead of in tubes. By providing communication between signalmen, the telegraph made a great contribution to railway safety. It enabled the old, haphazard time-interval method of regulating traffic to be replaced by the far more positive space-interval system known as 'absolute block'. For this purpose specialized single needle, three-position instruments were introduced, indicating 'Line Clear', 'Train on Line' and 'Line Blocked'. The blocked position was normal, the clear position being used to indicate that the signalman concerned was prepared to accept a train into his section.

Before this specialized system of railway telegraphy was introduced, however, telegraph lines had been extended beside many main lines by the Electric Telegraph Company (founded 1846) for purposes of general communication, the railway company paying a rental for their use with an option to purchase. By 1857 most of the larger towns in Britain had been joined in this way. There was also a telegraph link with France via a submarine cable which was successfully laid across the Channel at the third attempt in 1851. By 1862 there were 15,000 miles of telegraph line in Britain and a far greater mileage on the Continent and in America. In 1866, the monster *Great Eastern* at last succeeded in the epic struggle to lay a successful

Atlantic cable between Valentia in southern Ireland and Newfoundland, and by 1872 the Mayors of London and Adelaide were able to exchange messages.

In sharp contrast to the rapidity with which the telegraph system spread both nationally and internationally, at the purely local level telecommunication developed slowly. Until 1860, when two companies in London set out to remedy maters, it was still necessary to go to one of the offices of the Electric Telegraph Company in order to send a message to some distant town, the equivalent of today's trunk call. For the equivalent 'local call' there was still no alternative to the messenger boy. In 1859 the London District Telegraph Company was established to provide a local telegraph service with a central office at Charing Cross connected to a series of local offices within a four-mile radius. The company never achieved its target of a hundred branch offices because it was not conspicuously successful. The main reason for this comparative failure was that its Central Office never acted as a true 'exchange' by connecting two branch instruments together. This is inexplicable as it would have been technically perfectly feasible to have done so at this date. Instead, every message received at Charing Cross was laboriously decoded and passed to another telegraphist who recorded it for onwards transmission. This process was not only very tedious but tended to multiply errors. Consequently, the company's customers, on receiving an extremely belated and often garbled message, were apt to be cynical about Professor Wheatstone's much publicized statement that electrical impulses travelled at the rate of 288,000 miles a second. They felt that even a dawdling messenger boy would have been quicker and more reliable.

The Universal Private Telegraph Company founded in 1860, was more successful. It existed to provide private telegraph instruments and lines between related business offices and, unlike the unfortunate District Company, its activities were not confined to London but soon expanded to Glasgow, Manchester, Liverpool, Leeds and Newcastle. This idea had originated in 1857 when the enterprising brothers Sydney and Alfred Waterlow connected their City offices in London Wall and Birchin Lane by private overhead wire.

Because the cost of laying an underground line was found to be prohibitive the brothers obtained permission from property-owners to carry their wire over the roof-tops. For the same reason the two telegraph companies were forced to follow suit, being greatly helped by wires of improved strength introduced by W.E. Henley. This origin of the 'wire-scape' did not go unopposed. The Drapers Company refused the brothers Waterlow permission to carry their wire over the company's property by claiming that the freehold extended 'from the centre of the earth to the canopy of heaven: *Cujus est solum, ejus est usque ad coelum*'. An outraged gentleman wrote a letter to the press complaining thus:

> . . . the magnified linen posts and clothes' lines . . . render hideous our most public and best constructed streets . . . dwarfing the apparent altitude of some of our finest commercial architectural elevations.

What he would think of our modern urban skyline does not bear contemplating.

Both companies were eventually taken over by the new Postal Telegraphs Department of the General Post office in February 1870. They were soon to be superseded by the invention of the telephone by the Scot, Alexander Graham Bell. Bell demonstrated his invention to the British Association in 1877, the National Telephone Company was formed in the following year and Britain's first telephone exchange was operating in London in 1879. Like its predecessors, the company's activities were eventually taken over by the Post Office.

The final scientific research that made the use of electricity possible for lighting and power transmission purposes was carried out by Michael Faraday at the Royal Institution in a classic series of experiments on the relationship between electricity and magnetism that led to the discovery of the principle of electro-magnetic induction. In October 1821, he succeeded in making a circle of energized wire rotate about the pole of a magnet* and, alternatively, in making the magnet circulate round the wire. The electric motor had been invented. Exactly ten years later, in October 1831, Faraday's researches led him to make a second historic experiment. He caused a copper disc to rotate between the poles of a magnet and arranged electrical contacts on the rim of the disc to which wires were attached leading to a galvanometer. The latter registered current. In his historic paper *Experimental Researches in Electricity* which Faraday read before the Royal Society, he concluded his account of this experiment with these words: 'Here, therefore, was demonstrated the production of a permanent current of electricity by ordinary magnets.' Faraday had invented the dynamo and from this experiment stemmed a gigantic world-wide electrical power industry.

If the story of Sir Robert Peel's visit to the Royal Institution is true, Faraday must have had some inkling of the immense practical significance of his discovery. For when Peel noticed his experimental dynamo and enquired what its purpose was, Faraday is said to have replied: 'I know not, but I wager that one day your government will tax it.'

The translation of simple laboratory apparatus into practical working machines is often painfully slow and

---

*The wire floated on the surface of a bowl of mercury. The magnet was stuck upright in the centre of the bowl, its pole protruding above the surface. See L.P. Williams, *Michael Faraday*, p. 156. The subsequent experiment of 1831 is desccribed on p. 195.

particularly so where, as in this case, the working principle is entirely novel. The commercial production of reliable dynamos and electric motors involved the solution of many difficult technical problems such as the production of suitable insulated wire and the accurate winding of it into the coils of electro-magnets and armatures; the design and construction of a reliable commutator from copper segments insulated from each other proved a particularly difficult proposition.

Faraday's papers were widely circulated in Europe and the first attempts to give his discoveries practical shape emanated from France. The first of these, produced by Hippolyte Pixii in 1832, was a hand-turned generator with rotating magnets and fixed coils, but a dynamo demonstrated to the British Association at Cambridge shortly afterwards had the now orthodox arrangement of rotating coils and fixed magnets. Small generators of this type were being made in London from 1834 onwards.

All these early generators produced alernating current. This was held to be a great disadvantage by those for whom direct current from batteries had hitherto been the only source of electrical supply. Efforts were therefore devoted to the production of a direct current dynamo, with no inkling that the a.c. dynamo would become the great power supplier of the future.

In the 1850s fixed electro-magnets began to be used in dynamos, though permanent magnets were at first retained because they were considered essential for preliminary excitation. But in 1866 C.F. Varley made the important discovery that electro-magnets retained enough residual magnetism in their soft iron cores to make a dynamo so fitted self-exciting.

But it was not until 1870 that the true ancestor of the modern dynamo appeared in Paris. The invention of a Belgian engineer Théophile Gramme, this was a series wound d.c. generator with a commutator consisting of insulated copper segments. It was found that it would work equally well as a motor. With improvements made jointly by the Swiss engineer Emil Burgin and Colonel R.E.B. Crompton, the Gramme machine was developed in this country into the Crompton dynamo which, together with the somewhat similar generator developed by the brothers William and Werner Siemens, supplied current to most of the early installations in Britain.

Through the medium of carbon arc lamps, electricity was first used in lighthouses on the initiative of British and French engineers. Using a steam-driven dynamo with an output of 1½ kilowatts designed by the French engineer Baron A. de Meritens, Frederick Holmes gave a demonstration of carbon arc lighting to the Brethren of Trinity House in 1857. So impressed were the Brethren that the apparatus was successfully installed in the South Foreland lighthouse in December 1858,* while that at Dungeness was similarly equipped two years later.

In such early lighthouse installations the carbon electrodes were arranged horizontally, the arc being struck between their opposed tips. As the carbon slowly burnt away, some mechanism had to be introduced to advance the electrodes if the arc were not to grow dim and eventually fail. A clockwork mechanism was at first used, but later the english engineer, W.E. Staite, introduced a

---

* The sight inspired Charles Tennyson Turner to write a sonnet 'The South Foreland Electric Light' of which the opening lines run:

> From Calais pier I saw a brilliant sight,
> And from the sailor at my side besought
> The meaning of that fire, which pierced the night
> With lustre by the foaming billows caught.
> 'Tis the South Foreland!' I resumed my gaze
> With quicker pulse, thus, on the verge of France,
> To come on England's brightness in advance!

weight-driven mechanism whose speed was regulated by the expansion or contraction of a copper rod acted upon by the heat of the arc, an early application of the principle of the thermostat.

The type of arc lamp known as the Jablochkoff candle, because it was introduced by a Russian engineer of that name, overcame this difficulty. In it the carbon electrodes were opposed vertically and parallel with each other so that the gap between them remained constant as the carbon burned away.

It was the improved Gramme dynamo combined with the Jablochkoff candle that made possible Britain's first public electric lighting installations. The West India Docks were electrically lit, followed by Billingsgate Market, Holborn Viaduct and part of the Thames Embankment. The first industrial building in England to be so lit was a new pipe-casting foundry at the Stanton Ironworks in Nottinghamshire. Crompton was responsible for this installation, importing a Gramme dynamo and other equipment from France in 1877. He was not at this time an electrical engineer, having previously been chiefly concerned with steam road transport in India, but he did the work at the request of his cousins, George and John Crompton,* who owned the Ironworks. But its success was such that he was soon asked to carry out similar installations, notably at Whiteley's store in Bayswater. Finding, to his surprise, that he now enjoyed a considerable reputation as an electrical engineer, not only in England but in France, he decided to devote himself to this new and promising field and founded the firm of Crompton & Company, Arc Works, Chelmsford, to manufacture and install electrical equipment. It was here that the Gramme

---

* They were descendants of Samuel Crompton, the inventor of the spinning mule.

dynamo was significantly improved, most notably in the construction of the commutator.

The first important contract to be carried out by Crompton's new company was the lighting of Glasgow St Enoch station for the Glasgow & South Western Railway. In 1879 Crompton, in association with Messrs Marshalls, the agricultural engineers of Gainsborough, produced the first portable generating set by mounting one of his dynamos above the smoke-box of an ordinary agricultural portable engine and driving it by belt from the engine flywheel. This engine, ancestor of all showmen's engines, was brought to London at Christmas 1879, and parked in the Mews behind Crompton's house in Porchester Gardens. Throughout the following January it supplied current to a series of small arc lamps which lit Crompton's drawing- and dining-rooms. They were the first private rooms to be lit by electricity in Britain.

It was quite obvious that the carbon arc lamp would never be a suitable means of domestic lighting, but in the same year of 1879 this problem was solved. In response to an urgent request from his friend Joseph Swan, Crompton visited Swan's laboratory in Newcastle where he saw, burning brightly and steadily, a row of twenty small incandescent electric lamps. The idea of producing light in this way was not new, but the problems of exhausting the glass globe and of finding a suitable material for the filament had not been solved. Now Joseph Swan had provided the answers. He had produced a satisfactory filament from a cotton thread by partially converting it into cellulose by immersion in sulphuric acid and then carbonizing it. Unknown to each other, the American inventor Thomas Edison arrived at a similar solution by a somewhat different route in 1881 and had protected it by patents which effectually blocked the English inventor. However, in the meantime, Swan had patented some improvements so, after legal arguments, both decided that each had something to give the other and the Edison & Swan United Electric Light Company Ltd was founded in 1883.

As soon as he saw Swan's lamps, Crompton realized that the era of commercial electric lighting from public supply on the lines of the gas companies had now arrived and he urged Swan to commence manufacturing his lamps as quickly as possible. A display of Swan lamps supplied by a Crompton dynamo was first shown to the public at an exhibition in Glasgow and orders for the lighting of the Glasgow General Post Office and Queen Street station resulted. The several installations in London which followed culminated in the lighting of the Law Courts. This last was the largest installation of its kind in the world at the time of its completion early in 1883.

Although French engineers had first taken the initiative in the practical application of Michael Faraday's discoveries, by the 1880s British engineers had won a decisive lead over their Continental rivals. Not only did Crompton's company supply dynamos to Sulzer Brothers in Switzerland, but they carried out important lighting installations at the Royal Opera House and the Burg Theatre in Vienna, at Gothenburg and at Le Mans. But, as Crompton ruefully notes in his *Reminiscences*, his small pioneer undertaking failed to obtain the necessary financial backing from Britain. He appealed in vain to British merchant bankers for the support which would have enabled him to compete against Continental state-subsidized concerns and so was forced to retreat and confine his activities to England. Here, after lighting Tilbury and Barry Docks, Crompton was responsible for founding, in 1886, the Kensington & Knightsbridge Electric Supply Company, one of London's first public electricity suppliers. This had its origin in a modest scheme to supply current to new houses on the Kensington Court Estate and long after it had outgrown this small beginning it continued to be known as 'The Kensington Court Company'.

Such early generating stations supplied direct current via the medium of storage batteries and these were a very useful reservoir to tide over the inevitable early crises due to mechanical or electrical failures of the generator. In the case of the Vienna Royal Opera House, for example, the first theatre in the world to be electrically lit, the 400-volt storage batteries were in the theatre and were supplied with current from the generating station a mile away (which also supplied the Burg Theatre) by bare copper conductors supported on insulators within an underground culvert. At the Opera House, current was tapped off the batteries at 100 volts, this being the highest voltage for which Joseph Swan could then produce bulbs.

To the older Victorian generation, accustomed to gaslight, electricity in all its manifestations was quite incomprehensible. It was regarded with suspicion as some form of magic, and possibly black magic at that. It was believed that some new kind of gas flowed through the wires so that, if a bulb was removed from its socket and the switch were not turned off, the gas would escape. Similarly, in the humming of telegraph wires in the wind people believed that they could hear the murmur of voices. One indignant old lady wrote to the London District Telegraph Company to say that while she could tolerate the incessant babble of voices in the wires, she really must protest against the scandalous nature of the conversations. In another generation she would have been fair game for the psychiatrist.

The man who, above all others, deserves to be regarded as the father of our modern electrical engineering and supply industries was S.Z. de Ferranti. He foresaw that the future lay with the large generating station, distributing over a wide area. Unlike Winsor, however, whose plans might never have materilized without the technical assistance of Samuel Clegg, Sebastian Ziani de Ferranti was a brilliant engineer who was able to solve unaided all the formidable technical problems involved in the large-scale distribution of electrical power. 'Bold in his plans, but right', had been Sir Daniel Gooch's summing up of I.K. Brunel, and the same could be said of Ferranti. He was, indeed, the Brunel of electrical engineering and it is oddly appropriate that Ferranti's great Deptford Power Station should be built on the very site where the *Great Eastern* completed her fitting-out, looking across the river to Millwall, where Brunel had launched his mighty ship thirty years before. Both projects were comparative failures because the minds of their creators had moved too far ahead of their times.

Ferranti had been drawn to electrical engineering at a very early age and as a student at University College he had been fascinated by Crompton's early lighting installation at the Alexandra Palace where he met and talked with the engineer. Very soon after this, at the age of seventeen, he was lucky enough to be given a job by Crompton's great rivals, the Siemens Brothers, entering the Experimental Department of their works at Charlton. By the time he was nineteen he had already invented an a.c. generator, or 'alternator', of vastly improved efficiency, an arc lamp and a meter for measuring the consumption of electric current. This last was a vital instrument for public supply companies, and in 1883 he began manufacturing it on his own account in a small workshop in Hatton Garden.

It was at this point that, through the supply of his meters, Ferranti became associated with the Grosvenor Gallery Company. This, like Crompton's company in Kensington, had begun as a modest project to light the Grosvenor Gallery but had grown into a public supply company with an output of 1,000 kilowatts. Unlike most early generating systems, feeding it into overhead mains at a pressure of 1,200 volts, using simple transformers connected in series. The installation was far from satisfactory, and since there could be no storage batteries to tide over failures, there were numerous complaints from consumers. In their difficulty, the Company consulted Ferranti and were evidently so impressed with his ability that they appointed him Chief Engineer in 1886 when he was still only twenty-two years of age.

Such were Ferranti's energy and genius that in two years he had completely redesigned and re-equipped the station according to his own ideas. The pressure of the mains was doubled to 2,400 volts and the two new 750 h.p. alternators of his own design were installed in place of the earlier Siemens units. Transformers, also of Ferranti's design, were connected in parallel in order to reduce the pressure for domestic consumption. This installation was a small prototype of the modern system of power supply and it proved so successful that in August 1887 a new company, the London Electricity Supply Corporation Ltd, was founded to carry out the ambitious plans of the young engineer.

Ferranti must have had something of Brunel's magnetic personality to persuade hard-headed businessmen to back him in a scheme that seemed almost visionary in its magnitude. For he planned the widespread distribution of current on a scale and at a main pressure such as had never been contemplated before from a single large power station. Ferranti selected a Thames-side site at Deptford for his new station where there would be convenient supplies of water for cooling and boiler feed and also direct supplies of water-borne coal. He designed this Deptford station to have an ultimate capacity of 120,000 h.p., sufficient to light the whole of London at that time. Overhead mains would supply current at a pressure of 10,000 volts to a series of local sub-stations where it would be reduced to a street mains pressure of 2,400 volts. It would then be converted by small units in the cellars to the 100 volts then required for domestic lighting.

Wayleaves over the South Eastern Railway were obtained to carry this main supply into London and by October 1889, two Ferranti 1,000-kilowatt alternators were at work at Deptford, each driven by a vertical triple expansion engine with Corliss valve gear of 1,500 h.p. built by Hick, Hargreaves. Of these *Engineering* wrote: 'The new electrical machinery is so enormous as compared with anything in existence, that it may be deemed a perfectly novel creation.' Others called Ferranti the Michelangelo of Deptford because, as the *Electrical Engineer* wrote: 'from first to last, from foundation to top of highest turret, architecture, materials, foundations, and machines, all were specified or designed by one man.'

But Ferranti had even bigger things in mind for Deptford in the shape of two sets each of 5,000 h.p. Because these huge steam engines would revolve at comparatively low speed, the Ferranti alternators to which they would be directly coupled were of gigantic size. Their armature shafts were three feet in diameter and weighed seventy tons. They were two of the largest steel castings ever made and the Company had to install special plant in order to machine them. The finished armatures were forty-two feet in diameter. But, alas for Ferranti's hopes, these two enormous sets, each weighing 500 tons, were never installed but had to be abandoned when in an advanced stage of construction.

What became known as the 'Battle of the Systems' had developed between the London Electric Supply Corporation, representing Ferranti's advanced ideas, and the advocates of direct current supply from a series of small local generating stations.* The latter were helped by the public fear of a mains pressure of 10,000 volts, a fear which the L.E.S.C. endeavoured to allay by circulating a leaflet featuring a simple circuit diagram showing how pressure would be reduced for domestic consumption. The leaflet

---

* A similar battle was waged in the United States with George Westinghouse championing high-pressure a.c. against Thomas Edison's d.c. system.

quoted W.H. Preece, F.R.S., Engineer-in-Chief of the Post Office, as saying: 'The prejudice against High pressure is still strong; it is thought to be unsafe, but time and experience will eradicate this impression as they ultimately eradicate every fallacy.' But it was all in vain. The Electric Lighting Act of 1882 and its amendment of 1888, which permitted franchises of forty-two years to be awarded to public supply companies, were followed in 1889 by a Board of Trade inquiry to decide on the method to be used in the lighting of London and how allocations were to be made between rival companies. This ruled that contracts should be awarded on a parochial basis, local supply companies making their agreements with the respective parish vestries. The low-pressure d.c. system had won the day, making Ferranti's great power station at Deptford a costly white elephant. But although he lost the battle, he won the war, for his was the system which has ultimately prevailed all over the world.

Electricity generation set steam engineers new problems. For a steady output of current, particularly when driving alternators, absolutely constant speed was essential. The orthodox centrifugal governor, which James Watt had originally developed from a device used by the early millwrights to regulate their millstones, proved insufficiently sensitive and was replaced by a form of electrical governing. In the earliest installations the dynamo was belt-driven from the flywheel of a low-speed engine. Next, several types of high-speed steam engine were developed which could be directly coupled to the dynamo. Of these the Willans and the Bellis & Morcom engines were conspicuously successful. Both were vertical engines, totally enclosed.

Peter Willans was a friend of Crompton who had established a works at Thames Ditton to build a high-speed launch engine which he had designed. Crompton encouraged Willans to develop this engine into a larger unit suitable for power generation. This was very unorthodox in having two sets of triple expansion cylinders arranged vertically like the tiers of a wedding cake. It was single-acting, the steam passing from one cylinder to the other by means of a piston valve working within a hollow piston rod. Built in sizes up to 500 h.p., the Willans engine proved so successful that Peter Willans moved to a new and larger works at Rugby. Crompton's Vienna power station, for example, used six Willans engines, each of 150 h.p. and running at 500 r.p.m. In their heyday Willans engines totalling 53,340 h.p. were in use for generating current.

The rival engine built by Belliss & Morcom of Birmingham was more orthodox in design but possessed one unique feature of great historical significance. A pump in the sump circulated lubricating oil under pressure through ways drilled in the crankshaft to the main and connecting rod bearings. This forced lubrication system was the invention of Albert Pain, a draughtsman at the Belliss works, and it was destined to become the standard form of lubrication in all internal combustion engines in the next century.

Parsons turbo-alternators were first installed in the Forth Banks power station at Newcastle in 1892 and from this date the turbine rapidly ousted the reciprocating engine from the generating station. It was lower in first cost: it was more compact and required no massive foundations. Moreover, as generating sets increased in size in response to the growing demand for electrical power, the superiority of the steam turbine in economy and efficiency became the more marked. The original sets at Forth Banks had an output of seventy-five kilowatts, but by the end of the century 1,000 kilowatt Parsons sets were in use at Elberfield and already an output of 6,000 kilowatts was being talked about.

The possibilities of electricity, not only as a new means of lighting, but as a method of transmitting power to a distance was recognized and exploited as early as 1879. In that year there were two electric motors in use in Britain, one at Poynters' chemical works at Greenock and one driving a circular sawbench on Sir William Armstrong's country estate at Rothbury. The advantages of the electric motor were so obvious in so many power applications that from this small beginning their use spread very rapidly. In the more advanced factories they were adopted in place of steam engines for driving the plant. Later, it was recognized that the electric motor could be used to power individual machines, thus doing away with complicated systems of mechanical power transmission by rope drive or line shaft and belts. But the social significance of electrical power would not become apparent until the next century when the high-pressure a.c. system pioneered by Ferranti was generally adopted. For the first time the location of power-consuming industries need no longer be confined to the coalfields of the Midlands and the North where fuel was cheap. Industry forsook the area which had cradled it and followed the marching pylons into the southern counties of England to establish itself in towns which the impact of the Industrial Revolution had hitherto but lightly touched.

Electricity also offered a convenient new form of traction for railways. The idea of using some alternative to the steam locomotive which would transfer the source of power from the rails to the lineside had occurred to engineers in the earliest days of railways. But the only alternative to rope haulage they could then devise was the so-called 'pneumatic' or 'atmospheric' system by which trains were propelled either by air or atmospheric pressure acting upon a piston running in a continuous tube. One such system showed promising results and was tried out on quite a large scale,* but it was soon beset by technical troubles and thereafter the steam locomotive reigned supreme and

* See *Atmospheric Railways* by Charles Hadfield (Newton Abbot: David & Charles, 1967) for a full account of these abortive experiments.

unchallenged until the coming of electric power offered the best practicable alternative.

The possibilities of electric traction were first demonstrated by Werner Siemens at the Berlin Exhibition in 1879 where he showed a miniature electric locomotive of two h.p. picking up its current from a third rail. In 1883, in association with his brother William, Werner built the Giant's Causeway tramway in northern Ireland which was supplied with hydro-electric power. In August of the same year, Magnus Volk's pioneer tramway on the sea front at Brighton was formally opened by the Mayor. Both these railways were small-scale undertakings designed to exploit the novelty of electric traction. Nevertheless they performed a valuable service by demonstrating its practicability. That its use for street tramways or for underground railways would be particularly advantageous soon became apparent.

By this time Britain possessed over a thousand miles of urban street tramways. Most of these began by using horse traction but later many adopted steam locomotives. Owing to their use on the public roads as opposed to a reserved track, however, the steam tram had to comply with stringent Board of Trade requirements. Its speed had to be restricted by a mechanical governor; it must consume its own smoke and steam; it must have dual controls at either end and coupling and connecting rods must be covered by sheet-metal shrouding. Nevertheless, the steam locomotive was hardly a desirable vehicle to let loose in a street crowded with horse-drawn traffic, especially as the regulations were more often honoured in the breach than in the observance, being almost impossible to apply. Consequently electric power was eagerly seized upon as the best solution to the problem. The two pioneer electric tramways in Britain were those at Blackpool and Leeds. The former began running in 1885, picking up current from a third rail laid in a conduit beneath the road surface.

On its Roundhay Park route, electrified in 1891, the Leeds tramway used the system of pick-up from an overhead wire which was to become almost universal for street tramways.

The street tramway was one answer to the demand for more and better urban public transport that was created as the larger towns expanded ever farther into their surrounding countryside. But they were only a partial answer since they were affected by, and contributed to, traffic congestion in city centres. This had already become a problem in London by the 1840s. A better and, as time has shown, a more permanent solution was to construct urban railways underground.

The first underground railways in the world were London's Metropolitan and District lines which together now form the Inner Circle. Built with the object of connecting the various main-line termini, the first section to be completed was that part of the Metropolitan between Paddington and Victoria Street which was opened in 1863. This was not a true underground railway as we now understand the term because the greater part of it was constructed just below street level on the 'cut and cover' principle. This, as we shall see in the next chapter, proved an inordinately costly method, involving the diversion of all the cables, gas and water mains and sewers that now, like so many veins, arteries and alimentary canals, made one vast organism of the Victorian city. So the engineers resolved to build future underground railways at a much deeper level to avoid man-made obstacles. The first of these deep-level underground railways in the world was the City of London & Southwark Subway authorized in 1886. It was later decided to extend it to Clapham Common and to rename it the City & South London Railway.

On the Metropolitan and District lines, steam traction had been used from the beginning, but although the locomotives were fitted with condensers and were alleged to consume their own smoke, the atmosphere was apt to resemble the infernal regions. An intrepid reporter who made a footplate trip round the Inner Circle in 1893 wrote: 'The sensation altogether was much like the inhalation of gas preparatory to having a tooth drawn. . . . By the time we reached Gower Street I was coughing and spluttering like a boy with his first cigar.' 'It is a little unpleasant when you ain't used to it,' he quoted his driver as saying, 'but you ought to come on a hot summer day to get the real thing!'

It was obvious that steam traction would never do for a deep-level underground line, but the promoters of the City & South London began tunnelling without any clear idea what form of motive power they were going to adopt. There was talk of using a form of cable traction, but fortunately electric traction had by then just reached such a stage of development that the Chairman, Charles Gray Mott, could recommend its adoption without his sanity being questioned. In 1889 the contract for the necessary electrical equipment was awarded to Messrs Mather & Platt of Manchester who guaranteed that the working costs would not exceed 3½d. per train-mile.

The first section of the City & South London Railway between King William Street (now the Bank) and Stockwell was opened by the Prince of Wales on 4 November 1890. Initially, traffic was worked by small electric locomotives drawing current at 500 volts from a third rail, while a fourth rail provided the earth return. These locomotives drew trailer cars which became known to Londoners as 'the padded cells' because they had no windows. Their designers evidently argued that since the whole line was in tunnel, windows would be a superfluous luxury, but they overlooked that fact that it would be helpful to the passenger to see which station he had arrived at.

Although British engineers had been responsible for this pioneer electric railway, they failed to exploit their advantage. The initiative was lost to American engineers who had developed electric rail traction far more rapidly and on a much wider scale in the United States. The construction of further deep-level tube lines in London was largely American-financed and American-equipped. The same could be said of the subsequent electrification of the Metropolitan and District lines and also of many surface electric tramways. The present 600–volt d.c. system used on London Transport lines originated in America, and the ancestors of the present multiple-unit trains were imported complete from America for service on the Bakerloo line.

By the end of the century, the main-line railway companies had experienced a serious fall in their short-haul commuter traffic as a result of competition from street tramways and underground lines. Consequently they began to consider the electrification of their own suburban surface lines. Two companies which acted in this way early in the twentieth century were the London, Brighton & South Coast and the London & South Western Railways. Their electrified lines formed the nucleus of what has now become the extensive Southern Region system.

Soon, however, the railways wound find themselves facing a far more formidable competitor which, in the course of a few brief decades, would sweep thousands of miles of street tramways out of existence and with them that groaning galleon of the streets, the electric tram. But in the closing years of Victoria's reign the internal-combustion-engined vehicle was a feeble infant, the plaything of a few fanatics, not to be taken seriously.

One further application of electricity must be mentioned because it was destined to have a most profound influence upon the future of engineering. This was the production of aluminium by an electrolytic process developed independently by C.M. Hall in America and P.L.T. Heroult in France in 1886.

Aluminium had been produced in France by chemical means since 1850, but the cost of the process made it a luxury metal, the price never falling below 50s. a pound.

The new electrolytic process, however, promised abundant supplies at an economic cost provided an equally abundant supply of electricity was available cheaply, as it required approximately 20,000 kilowatt-hours of electricity to produce a ton of the metal. Because water power was the cheapest means of generating electricity, its availability determined the location of the first aluminium plants: at Schaffhausen Falls on the Rhine and at Niagara in America, while Switzerland, with her abundant water resources, was for many years the largest European producer.

By 1898 the British Aluminium Company was producing crude ingots by the Heroult process at Foynes from the alumina (aluminium oxide) extracted from the crude bauxite at Larne. These ingots were then sent to Milton in Staffordshire for further refining. Production was at first small, for even engineers could see no future for the metal other than int he making of domestic saucepans. However, the rapid twentieth-century development of the motor-car and the aeroplane as we know them would not have been possible but for the lightness and strength of aluminium and its alloys.

NINE

Civil Engineering After 1860

From 1860 onwards major civil engineering undertakings became increasingly the subject of highly organized teamwork between consulting engineers and specialist contractors. In Chapter 1 we saw how, in the first period of railway construction, a few great civil engineering contractors emerged out of the welter of small operators. These great contracting firms became so highly organized and employed such skilful and experienced engineers that an increasing amount of responsibility could safely be delegated to them by the employing company and by the consulting engineers concerned.

In his usual trenchant style, I.K. Brunel once defined a consulting engineer as a man who was prepared to sell his name but nothing more. This was certainly not true of the great consulting engineers of the second half of the century. They retained responsibility for the design of the works and were ultimately responsible to the promotors for their successful execution by the contractor. Nevertheless, it is significant that whereas the names of Stephenson, Locke and Brunel are widely remembered today, those of their successors – Barlow, Hawkshaw, Barry, Baker and Fowler – are largely unknown outside the ranks of their profession. In a more specialized and highly organized world they could not wield the same undisputed power and authority as did the engineers of the heroic age. In particular, the great contractor, by his rise to power, stole from the engineering consultant much of the limelight that was properly his due.* Because of this division of responsibility, this subtle shift in the balance of power in the engineer-contractor relationship, great civil engineering works after 1860 have been comparatively little celebrated because they lack the human drama of a single individual striving against odds. They were, however, very considerable achievements and worthy of record.

Although the combined length of the London Metropolitan and District lines and their branches amounted to only 17½ miles, so many unprecedented engineering difficulties were overcome in their construction that they represent the greatest railway-building feat of the post-1860 period. This is reflected in their total cost, which exceeded £10 million, and in the time – twenty-four years – taken to complete them. Construction began in March 1860 with the section of the Metropolitan between Paddington and Victoria Street (now Farringdon) under the direction of Sir John Fowler and the Inner Circle was finally opened in 1884 by the completion of the last section of the District line between the Tower and the Mansion House. For this last piece of line Sir John Hawkshaw and Sir John Wolfe Barry were jointly responsible, but the engineer most closely associated with the construction of the Metropolitan and District lines was Sir Benjamin Baker, first as Fowler's assistant and later as his partner.

Baker afterwards summed up the engineering experience which had been gained on this great work in these words:

It is now known what precautions are necessary to ensure the safety of valuable buildings near to the

* This process has been carried further in our own day. Some contractors now offer their customers a 'package deal' which includes design work, thus either eliminating the consulting engineer or architect completely, or relegating him to the position of a mere watch-dog on behalf of his client.

excavations; how to timber the cuttings securely and keep them clear of water without drawing the sand from under the foundations of adjoining houses; how to underpin walls and, if necessary, carry the railway under the houses and within a few inches of the kitchen floors without pulling down anything; how to drive tunnels, divert sewers over or under the railways, keep up the numerous gas and water mains, and maintain the road traffic when the railway is being carried underneath; and finally, how to construct the covered way so that building of any height and weight may be erected over the railway without risk of subsequent injury from settlement or vibration.

These lessons have stood modern civil engineers in good stead, particularly in the construction of new under-pass roads in urban districts, but in the nineteenth century it was experience dearly bought, although the cost per mile was progressively reduced as new techniques were perfected. Of the 13¼ miles of the Inner Circle, only just over 1,000 yards are in true tunnel; all the rest was built as open cutting using retaining walls or, in the case of the greater part of it, subsequently arched over. Although the line was laid out to follow the centre line of existing streets so as to minimize the disturbance to buildings and the disruption of road traffic, even so the engineers agreed that this 'cut-and-cover' method was much too costly and that any future urban railways must be built at deep level.

Although the deep-level tube line avoids all the engineering difficulties created by the cut-and-cover method, it presents problems of its own. Working at such depths, in order to ensure the safety of the tunnellers and to prevent what civil engineers term loss of ground, causing subsidence at the surface, it is necessary to employ a tunnelling shield.

The engineer responsible for inventing and using the world's first successful tunnelling shield was Sir Marc Brunel, the father of I.K. Brunel. He used it in the construction of the world's first sub-aqueous tunnel beneath the Thames from Rotherhithe to Wapping. Owing to the treacherous nature of the ground beneath the river, this proved to be a work of immense difficulty and hazard. Begun in 1825 it was not completed until 1843 when the young Queen bestowed a knighthood upon the engineer. Never was an honour more richly deserved, for without the tunnelling shield allied to the dauntless courage and tenacity of purpose of Marc Brunel, the work would never have been completed at all. Intended as a vehicular tunnel, the proposed spiral connecting carriageways were never built owing to lack of funds and this first Thames tunnel was used by foot passengers only until its adoption by the East London Railway in 1865.

The shield shown in Marc Brunel's patent specification was circular in form and afforded complete protection to the excavators working in its several compartments. It extended backwards in the form of a circular hood or diaphragm within which successive segments of cast-iron tunnel lining rings would be bolted together behind the advancing shield which was designed to be propelled forward by hydraulic jacks for which the last lining ring provided a purchase. Unfortunately, however, for various reasons, some technical but mostly financial, the actual shield first used in the Thames tunnel differed markedly from this. It was rectangular in shape, consisting of six triple-tiered sections, each of which could be advanced separately by screw jacks pressing against the end of the brickwork with which the tunnel was lined. Most important of all, it had no diaphragm, which meant that each time a section of the shield was advanced it left a small but vital section of the tunnel roof ahead of the brick lining exposed and unsupported. To this fatal defect many of the disastrous 'runs' and periodic inundations which occurred in the construction of the tunnel were due.

J.H. Greathead, the engineer chiefly responsible for the first London tube lines, designed for their construction a shield which followed much more closely Marc Brunel's original patent specification. It was circular in shape with a diaphragm within the protection of which the cast-iron tunnel lining segments were bolted up. It was also advanced hydraulically. It will be appreciated, however, that with this technique, as the shield moves forward, its diaphragm leaves behind it a void between excavation and lining. Greathead was able to fill this by injecting cement grout under pressure behind the lining rings through holes left in them for the purpose. In Marc Brunel's day such a technique had not been developed and, moreover, the decision to use a brick lining made the problem of the void much more acute. No doubt it was this difficulty which finally induced him reluctantly to forgo the protection of the diaphragm.

While Marc Brunel had to depend entirely on manual excavation, mechanical excavating machines were introduced by the builders of London tube lines. On the first to be built, the City & South London, the Thomson machine was used. This consisted of a series of small buckets with cutting edges strung on an endless chain mounted upon a movable arm which attacked the clay face through an aperture in the front of the shield. Later, a more efficient type of rotary excavator was introduced, the Greathead shield being adapted for the purpose by fitting to it an axial shaft and bearings on which the excavator arms rotated. This excavator, which was driven by an electric motor, was the invention of John Price of the contractors, Price & Reeves. It enabled a maximum tunnelling speed of 181 feet 8 inches a week to be achieved, a performance that was not surpassed when subsequent tube extensions were made in the 1930s.

The greater part of the tube tunnels were excavated through stiff London clay, but occasionally treacherous pockets of water-bearing sand and gravel were encountered such as had caused Marc Brunel so much trouble and anxiety. In this case, however, the difficulty was overcome by the use of compressed air in conjunction with air locks. Air at a pressure of 15–25 pounds per square inch effectually held back the water and prevented loss of ground.

J.H. Greathead died while the second tube line, the Waterloo & City, was under construction and his place was filled by Basil Mott on this and on the Central line which closely followed it, work starting in April 1896, at Chancery Lane. On all three of these railways Sir Benjamin Baker acted as consultant and, following the death of Sir John Fowler, Mott and Baker formed a consulting engineering partnership whose third member was David Hay, a member of the contractors's staff. As Messrs Mott, Hay & Anderson, the partnership exists to this day and still specializes in tunnelling work.

As previously noted, the first section of the City & South London was opened in November 1890. The short Waterloo & City, the first tube line to be built beneath the Thames, was formally opened by the Duke of Cambridge in July 1898. The main contractor was John Mowlem, another name that is still with us. The Central line from the Bank to Shepherd's Bush was opened by the Prince of Wales in June 1900.

Two major undertakings which occupied civil engineer throughout this period were the provision of water supplies to the growing towns and the construction of large floating docks, particularly in the Port of London, to accommodate larger ships and an ever-increasing volume of shipping.

The spectacular growth of the new industrial towns forced local authorities to look even farther afield in their quest for more water, a process that is still going on and which led to the construction of dams to impound water in suitable mountain valleys, often conveying it by pipe-line over considerable distances. It was in the new industrial towns bordering the Pennines that, not surprisingly, this need for water first became urgent. It was initially satisfied by forming a number of reservoirs in the Pennine valleys by constructing earth dams with a core of puddled clay. The technique of constructing such dams had been originally developed successfully by the great canal engineers of the eighteenth century who had to provide reservoirs to supply the summit levels of their canals. In suitable locations such earth dams are still built today, but in the nineteenth century when the theories of soil mechanics and hydraulics were in their infancy, as such dams grew larger and higher so the risk of disaster grew.

The first warning came in 1852 when a dam at Holmfirth, near Huddersfield, was undermined by leakage, fortunately without catastrophic consequences. It was in 1864 that disaster struck. The 100-feet-high Dale Dyke dam, one of a series constructed for the city of Sheffield, suddenly subsided when the reservoir had just been filled for the first time. The water overtopped the dam and in twenty minutes 80,000 cubic yards of earth and 200 million gallons of water had swept down the valley below. In the resulting devastation 250 lives were lost. Although evidence at the subsequent inquiry showed that the dam had been badly constructed, the precise cause of the initial subsidence was never established. It was believed to have been due to the collapse of a large cast-iron pipe beneath the superincumbent weight of the dam.

This disaster prejudiced engineers against the earth dam and subsequent nineteenth-century dams were massively built of masonry in accordance with a theory of construction propounded by the celebrated Professor Rankine of Glasgow. The first and most notable of these masonry dams was constructed at Vyrnwy in Wales to supply Liverpool with water. This was begun in 1881 with Thomas Hawksley as engineer-in-chief, but on his retirement in 1885 he was succeeded by George Deacon, the Liverpool waterworks engineer. Deacon had been in practical charge of the works since their commencement and was the true genius of the enterprise, using techniques of his own devising which anticipated modern methods in the use of concrete. The masonry used in the Vyrnwy dam was on a cyclopean scale, some of the individual stone blocks weighing as much as twelve tons. When completed in 1892 it was a fifth of a mile long and 144 feet high and it created what was then the largest artificial lake in Europe containing 12,000 million gallons of water. It is undoubtedly one of the most notable monuments to late Victorian civil engineering.

The example of Liverpool was soon followed by Manchester and Birmingham. Manchester's Thirlmere dam in the Lake District was completed in 1894, while Birmingham put in hand its gigantic scheme for a series of masonry dams in the Elan Valley in Radnorshire under the direction of James Mansergh, a celebrated specialist consultant on waterworks engineering who died in 1905, twelve months after the completion of the scheme. One of the Elan damns, that at Craig Coch, is 120 feet high. Apart from the building of the dams, the construction of the Elan piped aqueduct by which the water was conveyed through miles of hilly country to Birmingham was itself a considerable engineering feat.

Apart from a few small private docks, all shipping using the Port of London had to discharge or load in the tideway until the end of the eighteenth century. Small vessels could lie at riverside tidal quays, but larger ships such as the Indiamen anchored in the tideway and were serviced by lighters. The East and West India Docks on the Isle of Dogs, the first large enclosed docks in the port, were completed under the direction of William Jessop in 1802 and Telford's St Katherine's Dock followed in 1828. The construction of the Victoria Dock, 1¼ miles long, was begun in 1850 and opened by the Prince Consort in 1855. It was built by Peto, Brassey & Betts, the celebrated railway contractors, and it was noteworthy in that the Victoria Dock Company was the first in London to adopt hydraulic power on a large scale for operating the dock machinery.

Shipping increased in size so rapidly that by the 1870s these docks had become quite inadequate, even the Victoria Dock having an insufficient depth of water. Construction of the Albert Dock was therefore authorized in 1875 and work began under the direction of Sir Alexander Rendel. This was considered the finest and largest dock in the world at this time being 1¼ miles long and enclosing eighty-seven acres of water to a depth of twenty-seven feet. Despite its size, at £2 million it was the cheapest dock to be built and it was completed in a remarkably short time, being opened by the Duke of Connaught on behalf of the Queen in June, 1880. This low cost and speed was partly due to the discovery that the gravel excavated was suitable for making concrete, but partly also to the fact that it was one of the first major civil engineering projects on which mechanical methods were extensively employed. Ruston steam navvies, each capable of excavating 450–500 cubic yards a day, were used in placing the 500,000 cubic yards of concrete. Such machines soon disbanded the old mighty race of navvies whose unaided endurance and muscle power had built all the early railways. The Albert Dock contract was also noteworthy for being the first to be lit with electric light, which enabled work to continue after dark. It was said of the arc lamps that 'at no time would their light be less than that of a fine moonlight night'.

The 4,452-ton SS *Queen* of the National Line was then the largest ship using the Port of London and was the first ship to enter the new Albert Docks in 1880. The dock was later made capable of receiving ships of 12,000 tons but, despite this, so rapidly was the size of merchant shipping increasing that the decision was soon made to build new and larger docks at Tilbury.

The first sod of the new Tilbury Docks was cut on 8 July 1882 and they were completed in April 1886, when the SS *Glenfruin*, homeward-bound from China, became the first ship to enter them. Sir Donald Baynes, Bt, was the engineer of this project which, unlike the Albert Dock, was dogged by misfortune. Their cost, £2.8 million, was more than double the original estimate. The first contractors, Kirk & Randall, were dismissed as a result of a dispute and another firm, Lucas & Aird, appointed in their stead. This involved the dock company in five years of costly litigation. Finally, the luckless Tilbury company lost money heavily as a result of ruinous cut-throat competition with the London & St Katherine's Dock Company* which was finally brought to an end by amalgamation.

Sir John Fowler and Sir John Wolfe Barry also had a hand in the development of the Port of London. Fowler was responsible for the Millwall Dock, authorized in 1864 and built on 200 acres of marshland on the Isle of Dogs. The work included London's first large dry dock. It was built by the dock company and was 413 feet long. Frederick Duckham, the engineer of this dock company, was later responsible for developing and installing the first bulk grain-handling and storage facilities at Millwall in 1899.

Wolfe Barry superintended construction of the Surrey Docks in 1893. He described it as the most difficult job he had ever undertaken. There was so much trouble with running sand in the excavation that the dock works were not completed until 1904.

Unforeseen difficulties are something the civil engineer is always concerned to avoid. Not only can they constitute a hazard to the lives of the men under his command, but

* This company, was formed by amalgamation and owned the St Katherine's, East and West India and the Albert and Victoria docks.

they invariably empty the pockets of the shareholders in the company to whom he is responsible. Nevertheless, it is the courage, perseverance and resource with which such difficulties are met that transforms a routine job into an engineering epic. As *The Engineer* remarked of I.K. Brunel's efforts to launch his mighty ship the *Great Eastern*: 'A brave man struggling with adversity was, according to the ancients, a spectacle the Gods loved to look down upon.' It was precisely because they were venturing into the unknown, surmounting difficulties as they encountered them, that gives the work of the pioneer engineers this epic quality. It is also the fact that the engineer's growing expertise enabled him to anticipate and avoid such difficulties to a much greater degree that makes the work of later generations seem prosaic and unheroic. Efficiency is admirable from a financial point of view but it can be a soulless quality. It is in the nature of his profession, however, that despite every improved aid and access of knowledge, the civil engineer sometimes runs into difficulties which call out the old heroic qualities. So it was in the case of the building of the Severn Tunnel, a work which, in its difficulties and dangers and the determination and courage with which they were faced, is worthy to rank beside the greatest achievements of the pioneers. The tunnel was built under the direction of Sir John Hawkshaw, but the true hero of the story was the contractor, Thomas Walker.

Hawkshaw and Walker had previously been associated as engineer and contractor in constructing the connecting tunnels to enable Marc Brunel's Thames Tunnel to be used by the East London Railway. This was a difficult and delicate operation. Part of the tunnel had to be constructed in coffer dam directly beneath the London Dock and to avoid blocking the dock completely, this had to be built half at a time. A large sugar warehouse had to be supported on columns sixty feet high while the railway was carried beneath it. It was doubtless the efficiency with which Walker carried out this difficult assignment that led Hawkshaw to recommend to the Great Western Railway Company that he be awarded the contract for the Severn Tunnel. 'In my long experience of contractors extending over more than fifty years,' said Hawkshaw, 'I have never met with anyone surpassing Mr Walker for despatch in carrying out works.'

Thomas Walker had a lifetime of experience in railway construction for he began working on railway surveying in 1845 before he was seventeen. In 1847 he was employed by Brassey and worked for him, first on the construction of the North Staffordshire Railway and then for two years on the Grand Trunk Railway of Canada. He then left Brassey but worked for a further seven years in Canada on railway construction. Between 1863 and 1865 he was engaged on railway surveys in Russia, Egypt and the Sudan. He then returned to London where he managed construction work on the Metropolitan and District lines for the contractors Peto & Betts, John Kelk and the Waring Brothers. It was at this point that he decided to start a contracting business of his own and took up the East London Railway contract.

At this time the railway traveller between Paddington and South Wales was faced either with the long journey via Gloucester or with the shorter but more uncertain route which involved a ferry crossing of the Severn estuary between the timber piers of New Passage and Portskewett. The line of this ferry and its connecting railways had been determined by I.K. Brunel. The need for a rail-crossing of the Severn estuary had long been apparent to the Great Western directors, but such was the breadth of the estuary and so fierce its tidal currents that they were daunted by the magnitude of the task. It was not primarily the convenience of the passenger that they had in mind, but the need to provide a more direct route eastwards for the expanding coal traffic of south Wales.

The board supported a scheme for a new railway put forward by Sir John Fowler which included a viaduct 2¼ miles long and 100 feet above high water over the Severn at Oldbury Sands, but although construction was authorized in 1865, nothing was done and the scheme died in 1870. Also in 1865, a rival scheme for a tunnel beneath the estuary was advanced but was contemptuously dismissed. Its originator and chief advocate was Charles Richardson, an engineer who had originally worked under Marc Brunel on the Thames tunnel and subsequently as assiatant to I.K. Brunel on the original line of the Great Western Railway at Box tunnel. While Fowler's grandiose plan collapsed, the persistent Richardson managed to keep his despised tunnel scheme alive until it was eventually crowned with success when an Act for a Severn Tunnel Railway was passed in 1872. Further, Richardson succeed in securing the services of Sir John Hawkshaw as consulting engineer and by this means obtained the full support of the Great Western Railway Company.

The new railway was to be eight miles long from Pilning, on the line from Bristol to New Passage, to Rogiet in Monmouthshire (subsequently called Severn Tunnel Junction) where it would join the railway from Gloucester to South Wales. It included a tunnel just over 4½ miles long passing under the Severn on a line determined by Richardson at a point known as the English Stones. This proposed line intersected a deep channel of the river known as the Shoots, 400 yards wide and eighty feet deep. In order to ensure adequate rock cover, Richardson planned his tunnel to pass fifty feet beneath the bottom of the Shoots. This involved lengthy approach tunnels inclined on a gradient of 1 in 100 with the effect that little more than half the total length of the proposed tunnel would be under the river.

The Great Western Railway Company began work on the tunnel in March 1873, using direct labour, with

Richardson as engineer in charge and Hawkshaw as consultant. A shaft, subsequently referred to as the 'Old Shaft', was sunk at Sudbrook on the Monmouthshire side and the work of driving a seven-foot-square heading beneath the river from the bottom of this shaft was begun. This work was carried on in a singularly half-hearted manner. The labour force employed was small and, in 1877, after four years' work, only 1,600 yards of heading had been driven. At this stage tenders to complete the whole work were invited but, although three were received, all were rejected. Two small contracts were, however, let out. Oliver Norris, a local man, was commissioned to sink a shaft (called the Sea Wall Shaft) on the Gloucestershire shore and to commence headings eastwards and westwards from it. Rowland Brotherhood, a contractor who had done much work for the G.W.R. in the past, began sinking two further shafts, known as the Marsh and Hill Shafts, on the line of the tunnel west of Sudbrook, driving headings in each direction from them. Over these shafts temporary pumps were erected, but at Sudbrook a second shaft, known from its iron lining as the Iron Shaft, was sunk as a permanent pumping shaft and over it two Bull engines with cylinders fifty inches in diameter by ten feet stroke were erected.*

For the next two years, work went on more briskly and the heading under the river from Sudbrook Old Shaft had, by October, 1879, reached a length of nearly two miles and was within 130 yards of that being driven to meet it by Norris from the Sea Wall Shaft. Meanwhile a heading was commenced in a westerly direction from

---

* These engines were non-rotative and had no beams, the cylinder being mounted directly above the twenty-six-inch plunger pump. They were so called after Edward Bull who originated the type in Cornwall in an unsuccessful attempt to evade the Watt patent.

Sudbrook Old Shaft to meet that being driven towards it by Brotherhood's men working from the Marsh Shaft. It must be appreciated that the heading beneath the river was driven at a depth equal to the lowest level of the tunnel, i.e. fifty feet below the Shoots, whereas the entrance to this new western heading stood forty feet above it in the Old Shaft.

Such sub-aqueous tunnelling, where a shield cannot be used owing to the rock formation, is always fraught with the risk that, if an unexpected fault in the strata is encountered, the works may be instantly inundated from the river above. The irony of the story of the Severn Tunnel was that this peril came, not from the portion beneath the river, but from that section immediately westwards of it where the rock strata were extremely variable, fissured and faulted. But it may be that Richardson, after his early experience in the Thames Tunnel, was so obsessed with the hazards of sub-aqueous tunnelling that he paid insufficient attention to possible dangers in the approach tunnels in setting out his line. It was not a question of sinking trial shafts; even local knowledge could have alerted him to possible dangers lurking beneath the ground on the Monmouthshire shore had he troubled to inquire. The local inhabitants knew that in summer the little river Neddern, which rises in the hills above Llanfair Discoed, dried out near their foot only to burst out again near Caerwent at a point known as Whirly Holes. Such local evidence pointed to the existence of large subterranean fissures in the area carrying powerful streams of water. Sure enough, on 16 October 1879, the miners driving the heading westwards from Sudbrook Old Shaft broke into such a fissure at a point 300 yards west of the shaft. At once a cataract of water swept down the heading and thundered forty feet down the Old Shaft into the heading under the river which it soon completely filled. Five miles of the river Neddern and many local wells dried up. Fortunately no lives were lost in this disaster. The men working beneath the Severn dropped their tools, ran for their lives and were rescued via the Iron Shaft at Sudbrook.

By a further irony this inunduation coincided with the ceremonial opening of the Severn Bridge (see Chapter 1) which was attended by Sir Daniel Gooch, Chairman of the G.W.R. Gooch, in his speech at the luncheon, invited the guests to visit the Severn tunnel works. 'It will be rather wet,' he warned them, 'and you had better bring your umbrellas.' 'Alas,' was Thomas Walker's comment, 'he little knew how wet it was.'

It was at this juncture that Richardson retired from the scene and Sir John Hawkshaw assumed full control upon the understanding that Thomas Walker be given the contract to complete the whole of the works. He also stipulated that the level of the tunnel beneath the Shoots should be lowered by fifteen feet. Because the inundation had come from the land and not from the river, it is difficult to appreciate the reason for this decision. It caused a great deal of extra work, expenditure and time. Three shafts had to be deepened and a new drainage heading driven beneath the river at the new low level. It also meant that the approach tunnels had to be re-aligned. That on the western side was steepened to 1 in 90 but that on the east was maintained at 1 in 100 in the interests of the anticipated heavy eastbound coal traffic.

When Thomas Walker arrived on the scene he was greeted by a picture of desolation. The hub of activity at Sudbrook had been abandoned; the two Bull engines above the Iron Shaft and the forty-one-inch Cornish engine which had been installed temporarily over the Old Shaft stood cold and silent; the water lipped the tops of the shafts. It was immediately clear to Walker that in order to retrieve the drowned works a great deal more pumping power would be required. Two additional pumping engines were therefore ordered from Messrs Harveys of Hayle, Cornwall, and installed at Sudbrook. They were a seventy-five-inch Cornish-type engine with a thirty-eight-inch plunger pump and a seventy-inch engine with a twenty-eight-inch bucket pump.

It must have seemed to Walker as though there was a jinx on the works, for these Cornish pumps, usually among the most reliable machines in the world, gave him a great deal of trouble. First, the twenty-six inch pump of one of the Bull engines broke and before the divers could repair it the water level in the shafts had to be lowered. To do this the big Cornish engine with the thirty-eight-inch plunger pump was started up. At 7 a.m. on the morning of 2 July 1880 this pump burst and a man who had been in the pump well emerged white and shaking after a piece of iron casting eighteen inches across had whizzed past his ear. Suddenly relieved of their load, the pump rods ran out and the twenty-three-ton beam of the engine came down on its emergency catch wings or spring beams with a tremendous crash. Before it could be stopped the engine made another stroke and the beam 'came into the house' with another thunderous blow. Until it was scrapped recently, this engine bore witness to this frightening mishap in the shape of a crack in the massive cast-iron entablature supporting the beam trunnions.

The cause of the failure was traced to a new type of pump valve fitted by the makers which had failed to open, but it was not until 14 October that the big engine was repaired and set to work again. After 28½ hours' pumping it had cleared the Iron Shaft of water so that the twenty-six-inch pump could be repaired.

Next followed a most heroic episode. In order to facilitate the pumping-out operation, it became necessary to isolate the eastern portion of the drowned heading under the river. This involved closing a steel flood door 340 yards from the shaft bottom, tearing up the tramway rails which prevented its closure, shutting an eighteen-inch flap valve, and screwing down a twelve-inch sluice valve. A diver named Lambert volunteered to undertake this hazardous task. Followed by two other divers to keep his vital air hose clear of obstructions, Lambert picked his way through the darkness of a drowned tunnel littered with abandoned skips, tools, timber beams and lumps of rock. Despite the efforts of his companions, he was forced to give up after 270 yards because he was unable to drag the heavy air line any farther. The retreat was even more perilous since it was difficult to keep the air line free from kinks and coils among so many unseen obstructions. One such kink could have cut off the air supply on which Lambert's life depended.

This attempt was made on 3 November. Meanwhile Walker had heard that a diver named Fleuss was demonstrating at the Westminster Aquarium a self-contained diving suit of his own invention, with a compressed air supply carried in a knapsack. The resourceful Walker lost no time in bringing Fleuss down to Sudbrook with his patent suit. Accompanied by Lambert, Fleuss descended the Old Shaft, but on seeing the mouth of the flooded heading the inventor's courage failed him and he came out of the shaft saying he would not undertake such a task for £10,000. On 8 November, Lambert made a second attempt wearing the Fleuss suit but once again he failed. Two days later, however, the heroic Lambert entered the heading for the third time, and on this occasion he succeeded in lifting the rails, closing the steel door and operating the two valves.

Even so it took longer than had been expected to pump out the water and it was not until 27 December that it was possible to approach the door which Lambert had closed. Then the reason why the water had taken so long to clear was made plain. Lambert had done his work precisely as instructed, but unknown to anyone the twelve-inch stop valve which he had been directed to close had a left-hand thread and was already closed. Consequently Lambert had opened it fully when he thought he was closing it.

On 13 December, Walker and two companions clad in diving suits but wearing sou'westers instead of helmets set out to wade up the western heading to investigate the original cause of all the trouble – the fissure or 'great spring' as it was called. Walker determined to build two masonry headwalls fitted with water-tight doors to shut off the great spring from the eastward workings. Bricks and cement were loaded on rafts and hauled up the heading by ropes.

These headwalls were successfully completed by 4 January 1881, but a fortnight later it was the turn of the weather to add to Walker's difficulties. A great snow-storm buried Sudbrook under between three and four feet of snow. For four days no coal could be brought to the site and pumping had to stop. Walker himself was travelling down from London by the 'Zulu' express from Paddington, a famous broad-gauge flyer, on the day of the snowstorm. The train could not get beyond Swindon and Walker finally arrived at Sudbrook eighteen hours late.

Hawkshaw's decision to increase the depth of the tunnel meant that, from a constructional point of view, the original heading under the river became useless. As ventilation difficulties increased (there were several deaths from pulmonatory disease among the miners), Walker installed a Guibal fan seventeen feet in diameter at Sudbrook, exhausting the air through this old heading which was lined with brickwork and eventually became part of the permanent ventilation system. Meanwhile the work of driving a new heading beneath it was begun.

At the end of April, 1881, water broke into the heading under the river from the Seawall Shaft on the Gloucestershire side. The whole of the Salmon Pool, an area of water retained at low tide by the English Stones, drained into the workings. The fault was found and plugged with bags of clay, the method Marc Brunel had first employed to seal the bed of the Thames. Apart from one rock fall, this was the only trouble experienced in the underwater section of the tunnel.

After this difficulty had been overcome, the work went forward rapidly. Walker installed electric light in the workings, supplied from generators on both banks, and established telephonic communication between Sudbrook and Seawall. By October, 1881, workmen had succeeded in sealing the fissure through which the great spring had burst, while the new heading beneath the river extended for 1¾ miles. Then, on the morning of 10 October, disaster struck again. The great spring broke out at a new point with such sudden violence that the workmen in the vicinity were swept out of the workings before they could close the door in the headwall. Its flow was estimated at 27,000 gallons per minute and Walker, who had been hastily summoned, arrived at Sudbrook to see a river of clay water flowing down the tunnel invert and falling forty feet down the Old Shaft with a thunderous roar. Providentially, all those below contrived to make their escape and no lives were lost, but the workings were again completely drowned.

Once more Walker had to rely upon the courage of Diver Lambert to save the day. Wearing the patent Fleuss diving suit, on 29 October he made a gallant but unsuccessful attempt to close the door in the headwall. Refusing to admit defeat, Lambert announced that he would make a second attempt the next day, using an orthodox suit with two assistant divers as before to handle his air line. This perilous venture succeeded; Lambert closed the door, groped his way back to safety and once again the big Cornish engines swung to their work of clearing the flooded tunnel.

After this experience Hawkshaw and Walker agreed to make no second attempt to seal off the great spring. Instead, they determined to divert it into a new drainage heading driven beside the tunnel and leading into a sump beneath a third shaft which was to be sunk at Sudbrook. Over the mouth of this new shaft a permanent pumping installation of six seventy-inch Harvey beam engines would be provided. Three of these engines continued to work until November 1962, when they were superseded by electric submersible pumps.

But Walker's troubles were not at an end. Almost exactly a year later disaster struck yet again and from a new and totally unexpected quarter. The Severn estuary is noted for its exceptionally high tides. Spring tides rise thirty-two feet to produce the famous Severn Bore in the upper reaches, but the tide of 17 October 1883 was ten feet above the highest spring tide. Men called it a tidal wave, for such a tide had never been experienced before in living memory. A wall of water five to six feet high advanced over the low-lying lands on the Monmouthshire shore. It swept through workers' cottages and pumping stations, extinguishing the boiler fires; it overtopped the Marsh Shaft and plunged into the workings 100 feet below. The eighty-three men who were working in the tunnel retreated up the westerly gradient before the advancing waters until they could retreat no farther and turned to face death by slow drowning. They were rescued by their comrades in a boat which was lowered down the shaft. It must have seemed to Walker at this juncture as though all the forces of nature were conspiring to defeat him, yet with stubborn courage he fought back by adding to his mechanical armoury. In a surprisingly short time, in response to his urgent demand to Harveys of Hayle for more pumping power, four additional Cornish beam engines, two with seventy-inch and two with sixty-inch cylinders, were installed and set to work at the Marsh Shaft. These were second-hand engines which had seen service in the Cornish mines. To guard against any possible recurrence of such a tide he caused earth dykes to be thrown up to protect the Marsh Shaft and also the rim of the approach cutting at Rogiet.

Having thus successfully retrieved his command for the third time, Walker launched his final attack with a force of 3,628 men, the largest number ever employed on the tunnel, and by the end of 1884, the tunnel was complete with the exception of 300 yards in the vicinity of the great spring and 500 yards of invert in the deepest section beneath the Shoots. In October of this year the waters of the great spring had been successfully diverted into the new drainage heading and in 1886 the new shaft and permanent pumping station at Sudbrook were completed and brought into action.*

On 5 September 1885, 12½ years after work had begun, Sir Daniel and Lady Gooch with a party of friends were able to travel through the Severn tunnel in a special train. The opening of the tunnel for regular traffic was delayed, however, pending completion of the permanent pumping station and the installation of a new and larger Guibal fan for ventilation. It was finally opened for goods traffic on 1 December 1886.

Looking back on this stubborn civil engineering battle, Walker commented: 'Sub-aqueous tunnels have recently become quite the fashion. One such experience as the Severn Tunnel, with its ever-varying and strangely contorted strata, and the dangers from floods above and floods below, has been sufficient for me. One sub-aqueous tunnel is quite enough for a lifetime.' So saying he turned his back upon the Severn to face his last great undertaking – the construction of the Manchester Ship Canal. He did not live to complete it.

The importance of the part played by the Lancashire cotton industry and its capital of Manchester in Britain's

---

* The original two Bull engines and seventy-five-inch Cornish engine over the Iron Shaft were retained as were two forty-one-inch engines over the Sea Wall Shaft. Altogether there were fourteen pumping engines on the Severn Tunnel capable of raising thirty-four million gallons a day. All were in regular use until 1954.

Industrial Revolution cannot be overstressed. As we have seen, its complex machinery and insatiable demand for power brought the mechanical engineering industry into being. Its demand for better and cheaper transport created Britain's first canal, the Bridgewater, and, when this became inadequate, Britain's first railway of modern form, the Liverpool & Manchester. The fortunes of the cotton industry depended to a unique degree on efficient transport because it relied entirely on imported raw material and on its export trade. Consequently, by the last decades of the nineteenth century, even the railway would not satisfy it and a scheme which would enable sea-going merchant ships to reach Manchester began to be mooted.

Daniel Adamson, with the blessing of the Manchester Chamber of Commerce, was the original promoter of the idea of a ship canal and by 1882 two proposals had been put forward. Hamilton Fulton proposed to form a tidal canal by deepening and straightening the channels of the Mersey and Irwell up to Trafford Bridge, confining the Mersey between restraining walls in the upper estuary. This would have involved siting the proposed Manchester Docks in a considerable cutting and the scheme was abandoned in favour of the plan advanced by (Sir) E. Leader Williams. He advocated similar works in the Mersey estuary from Garston through Runcorn to Latchford, but a locked canal from the latter place, following the course of the two rivers and ascending by three locks to Manchester. This was the scheme which was brought before Parliament and defeated in 1883. It was submitted in a modified form once more in the following year only to be rejected again although Sir John Fowler and Sir Benjamin Baker both spoke in its favour.

The reason for this defeat, despite such powerful advocacy, was the equally powerful opposition organized by the city of Liverpool. No doubt this was motivated by possible loss of trade in Liverpool Docks, but it was argued cogently that the proposed alterations to the natural bed of the Mersey would reduce the scouring effect of the ebb tides and so greatly increase the rate of silting at the Mersey bar. This argument had the influential support of Sir Joseph Bazalgette, at this time President of the Institution of Civil Engineers, and of Captain Eads, an American engineer who had been responsible for opening up the mouth of the Mississippi. Eads enjoyed a world-wide reputation in this field and his appearance as an expert witness on behalf of the Mersey Docks & Harbour Board proved decisive. He maintained that to avoid harmful interference with the estuary, the proposed canal should be extended along the Cheshire shore below Runcorn to enter the Mersey in deep water at Eastham.

Having suffered two defeats, the promoters decided to take their cue from the opposition. Leader Williams prepared a revised scheme on the lines proposed by Eads which was put before Parliament and received the Royal Assent in August 1885. So far the project had cost its promoters £350,000 in legal and parliamentary expenses alone. It was at this juncture that Thomas Walker contracted to build the canal for £5,750,000, while the Bridgewater Canal was purchased by the Manchester Ship Canal Company for £1,710,000. Leader Williams was appointed the Company's engineer with Sir Benjamin Baker as consultant and the work was inaugurated by Lord Egerton, a collateral descendant of the famous 'Canal Duke' of Bridgewater, who ceremoniously cut the first sod on 11 November 1887.

Just over two years later, Thomas Walker died. Since the death of his brother and partner Charles he had been in sole charge of his contracting business and on his passing it was dissolved. Thereafter construction of the canal proceeded, partly by direct labour under the command of Leader Williams and partly by the contractors, Jackson & Wills.

In 1890 disastrous flooding occurred, water pouring over the banks and dams to flood the partly excavated dry bed of the canal. The resourceful Leader Williams decided not to attempt to rid the channel of water so as to continue dry excavation. Instead he made a virtue out of adversity by employing floating dredgers to complete the excavation to its full depth.

Apart from this one episode, construction was not marked by any unprecedented event and the canal was completed throughout by December 1893, opened for traffic on 1 January 1894 and formally opened by Queen Victoria on 21 May following. It is significant that the old Queen should have made the new Ship Canal the object of one of her increasingly rare public appearances. It was, indeed, a civil engineering work which was then unprecedented in its scale and magnitude. No doubt the ceremony reminded the Queen poignantly of earlier and far happier occasions in days, now long past, when her dear Albert had stood at her side. How he would have enjoyed it all!

And how the world had changed! Unlike the railway engineers the Queen had known in her youth, the civil engineer of the 1890s could summon a formidable array of mechanical armour to his aid. Despite the fact that as many as 16,000 men were employed, the greatest concentration of steam power ever to be assembled on a single project was used in the building of the Manchester Ship Canal. The included fifty-eight Ruston steam navvies, ninety-seven steam excavators, two French and three German land dredgers, the latter capable of excavating 2,400 cubic yards of soil a day, 194 steam cranes, fifty-nine pile drivers, 212 steam pumps and 182 stationary steam engines. In addition, 173 locomotives and 6,300 wagons operated over 228 miles of temporary construction lines. Altogether, 53½ million cubic yards of earth and sandstone rock were excavated, while to pitch the canal banks and to construct lock chambers, bridge piers and abutments thousands of tons of stone were imported: Cornish granite from Penryn, Derbyshire limestone from Crich and Yorkshire stone from Bramley Fell. And when the works were in full swing the steam machinery consumed 10,000 tons of coal a day.

There are triplicate entrance locks of different sizes at Eastham and duplicate locks at Latchford, Irlam, Barton and Mode Wheel. Each of the lock-gates for the largest of these locks consumed 230 tons of greenheart timber and twenty tons of ironwork. The gates were moved by hydraulic machinery operating at a pressure of fifty atmospheres designed and built by Armstrong Mitchell. To regulate the waters of the Mersey, Irwell and Weaver which fed the canal, huge counter-balanced Stoney sluices built by Ransome & Rapier were installed to discharge surplus water into the tidal estuary. As originally built, the first twenty-one miles of the canal between Eastham and Latchford locks was designed to be half-tidal, but because this caused silting, modifications were later made which excluded tidal water altogether.

Apart from the canal works proper, to provide the necessary headway of seventy-five feet for shipping, five railways had to be diverted onto new lofty embankments leading to steel bridges spanning the canal, while seven swing road bridges were provided. Most of these bridges were built by Sir William Arrol, the contractor responsible for the Forth Bridge, using Siemens-Martin open-hearth steel for the girder work in conjunction with wrought-iron floor plates the better to resist corrosion. The swing bridge linking Manchester with the new docks at Salford was the heaviest in the country at the time of its construction.

But the most remarkable piece of engineering on the canal was the Barton Swing Aqueduct. Because the canal occupies the old bed of the river Irwell at Barton, the historic Bridgewater Canal aqueduct, the first to be built in this country, which crossed the river at this point, had to be demolished. Such a proceeding would doubtless arouse a storm of protest today, but now the new aqueduct, designed by Leader Williams to replace it, has itself become an engineering monument. It consists of one swinging span of steel pivoting upon cast-iron rollers on the axis of a single central pier of stone. The waters of the Bridgewater Canal are carried in the steel trough of the span which can be isolated by the operation of hydraulic doors. The complete span, including the water within it, weighs 1,600 tons, but a single hydraulic ram, rising within the central pier when the aqueduct is to be swung, relieves the iron rollers of half this weight. In addition, when the aqueduct is cleared for Bridgewater Canal traffic to pass by the opening of the hydraulic doors, U-shaped wedges, lined with rubber and actuated by four rams, rise from below to seal off the gaps between aqueduct and canal.

The opening of the Manchester Ship Canal makes a fitting conclusion to this chapter. Although it was not a work of such difficulty and hazards as the Severn Tunnel, in sheer magnitude it was the greatest civil engineering achievement of the Victorian age. It was also a work which would have been impossible without mechanical aids. Finally, it was destined to be the last great work for which 'King Cotton' was responsible. Although none realized it at the time, the long years of his supremacy were numbered. But the engineering industry he had once sponsored would survive his eclipse and new industries arise in his stead.

TEN

The Shape of Things to Come

From the early eighteenth-century days of Thomas Newcomen down to those of Sir Charles Parsons at the end of the nineteenth, British engineers had led the world in the development of steam power in all its manifestations. To such a story of brilliant and sustained leadership, that of the introduction of the internal combustion engine and the motor-car makes a sorry sequel. For it was almost entirely due to the pioneer work of German and French engineers: Lenoir, Otto, Diesel, Daimler and Benz; Panhard and Levassor, de Dion and Bouton. When one surveys the ranks of contemporary British engineers, one can single out only two who are worthy to rank with these continental pioneers, Herbert Akroyd Stuart and Frederick Lanchester.

The famous – or infamous – 'Red Flag Act' which, until its repeal in November 1896, effectually prohibited the use of self-propelled vehicles on British roads, is usually made the excuse for British inertia in this new field. But while this may help to explain our tardiness where the motor-car was concerned, it cannot excuse our earlier inertia, our failure to recognize the boundless possibilities of a new prime mover. After all, the Frenchman, Etienne Lenoir, produced his first gas engine as a means of using town gas supply for power transmission as early as 1860, over thirty years before the first practical motor-cars appeared.

The fact is that British mechanical engineers rested too long and complacently on the laurels that the steam engine had won for them. They displayed a 'nothing like steam' mentality which regarded the pioneer efforts of continental engineers to develop a rival power with a mixture of contempt and suspicion.

Lenoir's gas engine did not employ the four-stroke cycle of suction, compression, explosion and exhaust which would later become orthodox in the internal combustion engine. It almost exactly resembled a simple single-cylinder horizontal steam engine. It even had a steam pattern slide-valve which admitted a charge of gas and air to each end of the cylinder alternately where it was fired by an electric spark. Because it did not compress the explosive mixture, Lenoir's engine was very inefficient by modern standards, but it was not for this reason that British engineers condemned it. They looked upon it as a travesty of a steam engine, for, to men trained in the steam tradition, the idea of substituting in the cylinder a violent explosion for the smooth and silent action of expanding steam was abhorrent. It seemed a crude violation of the almost sacred canons of mechanical principle which only a foreign engineer could be capable of perpetrating. This chauvinistic attitude was long-lived. After the Lenoir engine had been in existence for fifteen years, F.H. Crossley condemned it in these words: 'The common steam engine pistons of these engines, with their connecting rods and cranks, will not, under any circumstances conceivable by the writer, enable them economically to utilize the suddenly generated and suddenly expiring force of an explosion.* Ironically enough, however, in the very next year (1876) the German engineer, N.A. Otto, produced the first of his far more efficient horizontal gas engines working upon the four-stroke cycle and Crossley lost no time in acquiring a licence

---

* *Proceedings of the Institution of Mechanical Engineers*, 1875.

to manufacture the engine in Britain. It was so successful that after seventeen years 50,000 engines totalling 200,000 h.p. had been built by the firm of Otto and Langen in Germany and their licensees.

It was this Otto engine which Daimler and Benz would later make independent of the gas main by adapting it to burn 'light mineral oil' (petrol). This had hitherto been discarded as a waste product in the process of distilling paraffin from crude oil because it was too volatile and dangerous. The Otto gas engine was thus the true parent of the engines which power our modern motor-cars.

The ancestry of the high-speed oil engine is by no means so clearly established although it has brought about a revolution in our own day more far-reaching than that of the petrol engine. More economical and using a cheaper fuel than the petrol engine, it has proved a far more formidable rival of steam power, ousting it from commercial road transport, from the farmer's fields, from the engine rooms of all but the largest ships and banishing the steam locomotive from the railways of the world. Yet the name 'diesel', that is popularly applied to this engine and to the fuel it consumes, is a misnomer because in working principle it bears little resemblance to that of the engine patented in this country in 1892 by Dr Rudolph Diesel of Germany and successfully built by him at Augsburg at the end of 1895.

The oil engine utilizes the heat generated by the compression of air to ignite within the cylinder a finely atmoized spray of oil. To achieve this, Diesel pumped into the cylinder a blast of air at a pressure of 800–1,000 pounds per square inch along with the fuel spray, a method known as 'blast injection' which is now obsolete. The modern high-speed oil engine, on the other hand, relies solely on the heat generated by compression in the cylinder to produce combustion so that only fuel is injected into the combustion chamber. This working principle is known as 'solid injection' to distinguish it from the true diesel system. The undoubted pioneer of solid injection was Herbert Akroyd Stuart of Bletchley in Buckinghamshire whose oil engine was the subject of a series of patents taken out by him between 1886 and 1890. The type of engine invented by Stuart did not rely on the heat of compression alone to fire the atomized oil but used a pre-heated combustion chamber, and for this reason it is sometimes called a 'semi-diesel engine'. Nevertheless, in the design and operation of fuel pump and injector and in the way the explosion originated in a separate combustion chamber communicating with the cylinder through a narrow port, Stuart's engine is more closely akin to the modern oil engine than is Diesel's invention which it pre-dated. So Stuart has the stronger – though seldom acknowledged – claim to be its true father. The firm of Richard Hornsby and Sons of Grantham undertook the manufacture of Stuart's engine under the name of Hornsby-Akroyd. It proved popular, reliable and long-lived. Two of them, supplied in 1892 to a pumping station at Fenny Stratford, worked until 1932 and one is now preserved. The firm continued to build similar engines for over thirty years.

Despite the pioneer work of Akroyd Stuart and the important contributions made to the development of the oil engine by other British engineers in the first three decades of the twentieth century, German engineers and manufacturers have acquired a reputation for supremacy in this field similar to that held by British steam engineers in the nineteenth century. That so many modern British oil engines should be of German design built under licence betrays either lost initiative or lack of encouragement due to loss of faith in British engine designers.

The advent of the motor-car in this country was heralded by the 'bicycle boom' of the 1880s. The bicycle was not affected by existing highway legislation and it was relatively cheap. It brought to thousands a new mobility and the rediscovery of English roads which, apart from local horse-drawn traffic, had lain deserted throughout the railway age. The bicycle may be said to have prepared the way for the motor-cycle and motor-car in more senses than one, for many of the manufacturing techniques developed in the production of bicycles would be later adopted by the makers of their self-propelled successors.

Two key inventions transformed the primitive 'hobby-horse' and 'bone-shaker' into the modern bicycle. They were the wire-spoked tension wheel patented by Edward A. Cowper in 1868 and the pneumatic tyre patented in 1888 by John Boyd Dunlop.* Bicycle manufacture became concentrated in Coventry where the decline of the silk-weaving and clock-and watch-making trades had left a suitable pool of skilled labour. James Starley was responsible for introducing the new trade to the city and by 1885 there were no less than 170 firms making bicycles in Coventry alone. It is not surprising, therefore, that Coventry became the birthplace of the British motor industry although that birth was scarcely auspicious.

The first British-built cars were either inferior copies of continental makes, or French or German designs built under licence and making use of imported components. The only shining exceptions were the cars designed and built by Frederick Lanchester. At a time when French and German cars revealed only too clearly that they were adaptations of existing bicycle or horse-carriage practice, Lanchester was the first engineer to realize that such a dependence on past traditional practices was wrong and that the new vehicle

---

* Both were 're-inventions'. Sir George Cayley designed the first tension wheel in history in 1808. It was intended for an aircraft undercarriage. R.W. Thompson had fitted a horse-drawn brougham with pneumatic tyres as early as 1846. The tyres were of leather with inner tubes of rubberized fabric. Later, Dunlop insisted that his tyres were not suitable for motor vehicles and it was left to the French Michelin brothers to prove him wrong.

called for an entirely original approach. He built the first of three experimental vehicles in 1895 and, after five years of experiment, he founded his first small works in Ladywood Road, Birmingham. Here the first Lanchester production cars were turned out, their appearance coinciding with the Queen's death in 1901.

These first truly British production cars embodied a number of advanced design features which would later become commonplace. They had an engine that was properly balanced both statically and dynamically, an epicyclic 'easy-change' gearbox, drive by propeller shaft to a worm-drive 'live' rear axle, roller bearings and splined shafts in the gearbox. More than this, long before the word ergonomics had been coined, Lanchester made the convenience and comfort of the driver his first consideration. Because his car was literally designed round the driver, he built it complete with its body whereas the usual practice of the period and for some years after was to build a chassis only and to leave it to traditional coachbuilders to erect a body upon it on conventional horse-carriage lines.

Lanchester's true stature as an automobile engineer has only lately been recognized. Like Brunel and Ferranti before him, he trespassed too far into the future and paid the penalty. Although his outstanding talent and achievements were appreciated and honoured by the engineering profession in his lifetime, he never received any financial backing and his cars, though they soon won a high reputation among the more knowledgeable, only enjoyed a modest success. He died a poor man in 1946. In his old age he wrote sadly: 'The present day world to me is like a gigantic lunatic asylum, but then I am a survival of the old solid, sane Victorian period, and I think that no one who lived then can have a very high opinion of the methods of today.' So far as the overall design of the motor-car was concerned Lanchester's ideas had little immediate influence. The conventional triumphed and it is only in the last twenty-five years that the motor-car can be said to have rid itself of the last vestiges of horse-carriage influence.

Besides his automobile engineering activities, Frederick Lanchester interested himself in aeronautics with brilliant results but with even greater lack of recognition. For despite the world leadership given to Britian in the science of flight by Sir George Cayley at the end of the eighteenth century, with but few exceptions British engineers ignored this great legacy. They regarded with contempt the French invention of the balloon and adopted a similar attitude towards attempts to build a successful heavier-than-air flying machine. Lanchester's experience shows that this attitude persisted until the end of the century although it should have been obvious that, with the new power of the petrol engine to aid them, the efforts that were being intensified in other countries must soon succeed. In fact it was a mere two years after the death of Queen Victoria that the first successful flight by a heavier-than-air machine was made by the Wright Brothers in America.

Lanchester began investigating the problem of flight in 1892 and in 1894 he constructed a series of model gliders which he catapulted from the upper windows of his house at Olton, near Birmingham. As a result, he formulated his great vortex or circulation theory of sustentation in flight which is now recognized as the foundation of the modern science of aerodynamics and of aerofoil theory. Yet when he presented it in the form of a paper to the Royal Society it was rejected and when, later, he published it in book form its value was first recognized in Germany.

Again, Lanchester realized in 1892 that successful heavier-than-air flight depended on the development of a petrol engine having an exceptionally high power-to-weight ratio. In 1894 he proposed to start a factory to build such an aero engine but was dissuaded by an engineering friend who told him that, if he did so, his reputation as a sane engineer would be ruined.

In 1895, in the course of experiments based on the theories of Clerk Maxwell and the earlier experimental work of the German scientist Heinrich Hertz, Ernest Rutherford succeeded in transmitting a radio signal over a distance of three miles from the Cavendish Laboratory at Cambridge. But, once again, British engineers failed to follow up this advantage and it was left to the Italian, Guglielmo Marconi, to pioneer both radio telegraphy and telephony outside the laboratory. At the very end of the century Marconi patented his invention and on 12 December 1901 he succeeded in transmitting a radio signal across the Atlantic from Ireland to Newfoundland. Three years later, Sir Alexander Fleming produced the first thermionic valve which not only made possible radio as we understand it but was also the basis of modern electronic engineering which would give us television, radar, computers, space travel and satellite communications.

More significant for the future than all these scientific discoveries, however, in 1895 Rutherford began at Cambridge his experimental investigation into the property of radioactivity. Had he realized at the time the ultimate consequence of these experiments for all mankind, he might well have stayed his hand. For they would lead inexorably to the explosion of the first atomic bomb just fifty years later on 16 July 1945.

Thus at the close of the Victorian Age the seeds of the modern world had already been planted and most of them had begun to put forth their first tentative shoots. In the same way, developments which we regard as typically Victorian, such as railways and gas light, owe their origin to an earlier age. But a significant difference is apparent. Whereas at the time Victoria came to the throne British engineers led the world, by the time she died they had been overtaken by others and, in their own country, they were playing a role subordinate to that of the scientist. The career of Michael Faraday marks the beginning of this ascendancy of science over engineering as the pace-setter of material progress and Rutherford and Fleming inherited Faraday's mantle. But in the task of translating laboratory apparatus into practical commercial form other countries led the way.

Some of the more superficial reasons why Britian lost faith in her engineers and so speedily lost the industrial leadership of the world have been advanced in an earlier chapter. But some more fundamental theory must be sought to explain why it was that the course of the Industrial Revolution in Britain in the nineteenth century so signally failed to fulfil the optimistic hopes of those who had poineered that revolution in the eighteenth, allowing a mood of doubt and disillusionment to prevail which distrusted all further technical innovation.

The fathers of the revolution believed with Adam Smith that the principle of division of labour was all-important. In the past, they argued, it was the application of this principle which had alone enabled man to rise above the brute beasts. Now, armed with the new powers that machines had given him, they saw man standing upon the threshold of an era that would bring peace, prosperity and happiness to all.

In the nineteenth century, however, this roseate vision of the future paled before the bleak doctrine of Malthus who held that populations would always tend to outgrow any increase in the means of subsistence. 'A man,' wrote Malthus, 'who is born into a world already possessed, if he cannot get subsistence from his parents, in whom he has a just demand, and if the society does not want his labour, has no claim of right to the smallest portion of food and, in fact, has no business to be where he is.' It was their ready acceptance of this philosophy that enabled a new generation of wealthy manufacturers and merchants to regard with complacency the miserable circumstances of the labouring poor as though they were divinely ordained. This attitude inevitably provoked the concept of the 'class war', that potent Marxist philosophy of hatred which provided the poor and oppressed with a seemingly logical materialistic creed. But now that this nineteenth-century struggle between the haves and have-nots has passed into history, we may wonder whether it did not obscure the real issue – the validity of the optimistic philosophy that gave birth to the Industrial Revolution.

Adam Smith correctly foresaw certain dangers in the application of the principle of the division of labour by the aid of machines, but unfortunately he failed to carry his reasoning far enough. As Francis Klingender* put it:

> Smith recognized that as a result of the division of labour, 'the labouring poor, that is, the great body of the people', whose working lives are reduced to the monotonous repetition of a few simple operations, would necessarily tend to become 'as stupid and ignorant as it is possible for a human creature to become'. He therefore demanded that the State should intervene to prevent this evil by providing universal education.

Now that the goal of universal education has been achieved, the flaw in this argument is becoming increasingly apparent so that it seems surprising that a mind so acute as Adam Smith's failed to foresee the inadequacy of his remedy. That despite repeated bribes of more money for less work, universal education would surely lead to a growing dissatisfaction among those educated with that 'monotonous repetition of a few simple

* *Art and the Industrial Revolution*, 1947, 1st edn, p. 27.

operations' which they were subsequently called upon to perform should have been obvious. So long as education remains liberal – and there is a grave danger* that it may not – it must ultimately lead to a drastic collision between the educated and 'the system'.

At a time when Adam Smith's enlightened contemporaries, men like Erasmus Darwin and his fellow-members of the Lunar Society, could range freely and delightedly over the whole field of human knowledge, he may be forgiven the false assumption that there would always be such a catholic and brilliant intellectual *élite* to preside over human affairs and to guide the Industrial Revolution on its promising but perilous course. He failed to take into account, however, the speed with which the principle of the division of labour would be applied to intellectual as well as to physical work. As a result, knowledge accumulated as rapidly as wealth but at the expense of wisdom and humanity. The individual human person declined in stature as he became less of a whole man and more of a specialized thinking machine. We have seen this process at work in the engineering profession throughout the second half of the nineteenth century, but it was equally true of every other activity. It produced a terrifying failure of the sense of individual responsibility and its corollary, an uneasy feeling that man was no longer in control of his destiny; that human affairs in general and the Industrial Revolution in particular had acquired an irresistible momentum of their own. In this situation we can but envy those great but misguided pioneers who felt that they had the controls of the machine of material

---

* The danger is that our modern scientific powers may be used to avoid this collision by producing a generation both mentally and physically adapted to suit 'the system', as envisaged by Aldous Huxley in his *Brave New World*.

progress so surely under their command that they could guide it where they would.

In 1803, William and Dorothy Wordsworth were walking near Wanlockhead in Scotland with Coleridge when they saw, for the first time, a beam pumping engine at work at a colliery. They were fascinated by the repetitive motion of the great wooden arch-head of the beam with its dependent chains, nodding over the mouth of the mine. Dorothy wrote in her Journal:

> There would be something in this object very striking in any place, as it was impossible not to invest the machine with some faculty of intellect; it seemed to have made the first step from brute matter to life and purpose, showing its progress by great power. William made a remark to this effect, and Coleridge observed that it was like a giant with one idea.*

If such acutely perceptive minds could discern in so crude a self-acting machine some faculty of intellect, what would they have to say of our computers or of our complex automatic machine control systems? But the machine's gain has been our loss, for we have endowed it only by impoverishing ourselves, by becoming as specialized in function as the machines we create. Under the influence of mechanistic philosophy we even think of ourselves as machines and, like them, become giants with one idea, showing our progress by great power.

The study of evolution and biology provides salutary object lessons on the fate of over-specialized species. They are salutary because man has acquired an increasing measure of control over his own evolutionary process.

---

* I am indebted to my friend Sir Gordon Russell for drawing my attention to this quotation.

Should he perish like the dinosaur beneath the weight of his own armour it will be by his own conscious act. For, unlike the dinosaur, he has wilfully created that armour. That is the essence of the human tragedy.

To avoid such an ultimate disaster we need to recapture the mood of optimistic zest and eager inquiry that animated the pioneers in the last decades of the eighteenth century. But whereas they directed their energies to the pursuit of knowledge and invention, we need to devote ours to breaking down the barriers between the isolated compartments of knowledge. The urgent need is for synthesis no matter how difficult such a goal may be. All man's greatest achievements in art, science, philosophy or invention have been the fruit of the sudden fusing together in the individual mind of two hitherto disparate ideas. We call such a fusion inspiration. The further knowledge is pursued in specialized depth, the more difficult this marriage of ideas becomes. Hence the history of nineteenth-century Britain in general and of Victorian engineering in particular, illustrates the gradual failure of inspiration accompanied by a growing feeling of doubt, uncertainty and misgivings. Until with humility we strive for this creative synthesis, not all the millions that are now spent on research can recover lost certitude and reanimate the spirit of the pioneers. Nor can individual man achieve a like stature, or feel, like his forefathers, that he is wholly responsible for his destiny.

Select Bibliography

GENERAL

Armytage, W.H.G., *A Social History of Engineering* (London: Faber & Faber 1961; 2nd edn 1967).

Derry, T.K. and Williams, T.I., *A Short History of Technology* (Oxford: The Clarendon Press, 1960).

Hobsbawm, E.J., *Industry and Empire, An Economic History of Britain since 1750* (London: Weidenfeld & Nicolson, 1968).

Klingender, Francis D., *Art and the Industrial Revolution* (London: Noel Carrington, 1947; new edn, ed. Sir Arthur Elton, London: Evelyn, Adams & Mackay, 1968).

Norrie, C.M., *Bridging the Years, A Short History of British Civil Engineering* (London: Arnold, 1956).

Raistrick, Arthur, *Quakers in Science and Industry* (first published 1950, reprinted, Newton Abbot: David & Charles, 1968).

Rolt, L.T.C., *Great Engineers* (London: Bell, 1962) (short biographies of Maudsley, Murray, Locke, Fowler, Baker, Crompton etc.).

——, *The Mechanicals, Portrait of a Profession* (London: Heinemann, 1967).

Smiles, S., *Industrial Biography* (London: John Murray, 1883; new edn, Newton Abbot: David & Charles, 1967).

SELECT BIBLIOGRAPHY

1. THE RAILWAY ENGINEERS

Barman, Christian, *An Introduction to Railway Architecture* (London: Art & Technics, 1950).

Coleman, Terry, *The Railway Navvies* (London: Hutchinson 1965; Penguin Books, 1968).

Ellis, Hamilton, *British Railway History*, 2 vols (London: Allen & Unwin, 1954, 1959).

MacDermot, E.T., *History of the Great Western Railway*, 2 vols (London: Great Western Railway, 1927, 1931).

Paar, H.W., *The Severn & Wye Railway* (Newton Abbot: David & Charles, 1963) (the Severn Bridge).

Rolt, L.T.C., *Isambard Kingdom Brunel* (London: Longmans, 1957) Part 2).

——, *George and Robert Stephenson* (London: Longmans, 1960).

Simmons, Jack, *The Railways of Britain* (London: Routledge & Kegan Paul, 1961).

Sutherland, R.J.M., 'The Introduction of Structural Wrought Iron', *Newcomen Society Transactions*, Vol. XXXVI, 1963–4.

2. MEN OF STEAM

Barton, D.B., *The Cornish Beam Engine* (Truro: Barton, 1965).

Dickinson, H.W. and Titley, A., *Richard Trevithick* (Cambridge: C.U.P., 1934).

——, *A Short History of the Steam Engine* (Cambridge: C.U.P. 1938; 2nd edn, London: Cass, 1963) (includes history of the stream turbine).

Nock, O.S., *Steam Locomotive* (London: Allen & Unwin, 1957; new edn, 1968).

Rolt, L.T.C., *Thomas Newcomen, the Prehistory of the Steam Engine* (Dawlish: David & Charles, 1963).

——, *James Watt* (London: Batsford, 1962).

Watkins, G.M., 'The Vertical Winding Engines of Durham', *Newcomen Society Transactions*, Vol. XXIX, 1953–5.

3. SMOKE OVER THE SEA

Burtt, Frank, *Cross Channel and Coastal Paddle Steamers* (London: Richard Tilling, 1934).

Chaloner, W.H., 'Aaron Manby, Builder of the First Iron Steamship', *Newcomen Society Transactions*, Vol. XXIX,, 1953–5.

Duckworth and Langmuir, *Clyde and Other Coastal Steamers* (Glasgow: Brown Ferguson, 1939).

Rolt, L.T.C., *Isambard Kingdom Brunel* (London: Longmans, 1957) (Part 3).

Spratt, H. Philip, *The British of the Steam Boat* (London: Charles Griffin, 1958).

4. THE ENGINEER AND THE FARMER

Clark, Ronald H., *The Development of the English Traction Engine* (Norwich: Goose, 1960).

——, *Savages Limited, A Short History* (Norwich: Modern Press, n.d.)

Smith, Eng. Capt. E.D., 'Some Pioneers of Refrigeration', *Newcomen Society Transactions*, Vol. XXIII, 1942–3.

5. THE WORKSHOP OF THE WORLD

Gale, W.KV., *The Black Country Iron Industry* (London: The Iron & Steel Institute, 1966).

Halstead, P.E., 'The Early History of Portland Cement', *Newcomen Society Transactions*, Vol. XXXIV, 1961–2.

Hamilton, S.B. 'Sixty Glorious Years, the Impact of Engineering on Society in the Reign of Queen Victoria', *Newcomen Society Transactions*, Vol. XXXI, 1957–9 (public works engineering).

Henry Maudslay and Maudslay Sons & Field (London: The Maudsley Society, 1949).

Johnson, H.R. and Skempton, A.W., 'William Strutt's Cotton Mills, 1793–1812', *Newcomen Society Transactions*, Vol. XXX, 1955–7 (early iron-framed buildings).

McNeil, Ian, 'Hydraulic Power Transmissions', *Engineering Heritage* (London: Heinemann, 1963).

Nasmyth, J., *Autobiography*, ed. Smiles, S. (London: John Murray, 1883).

Pike, E. Royston, *Human Documents of the Industrial Revolution* (London: Allen & Unwin, 1966).

Rolt, L.T.C., *Tools for the Job, A Short History of Machine Tools* (London: Batsford, 1965).

Skempton, A.W., 'Portland Cements, 1843–87', *Newcomen Society Transactions*, Vol. XXXV, 1962–3.

6. HIGH NOON IN HYDE PARK

Great Exhibition of 1851, The, A commemorative album compiled by the Victoria & Albert Museum (London: H.M.S.O., 1950).

Hobhouse, Christopher, *1851 and the Crystal Palace* (London: John Murray, 1950).

Skempton, A.W., 'The Boat Store, Sheerness and its Place in Structural History', *Newcomen Society Transactions*, Vol. XXXII, 1959–60.

7. THE AGE OF STEEL

Gale, W.K.V., *The British Iron and Steel Industry* (Newton Abbot: David & Charles, 1967).

Osborn, Fred M., *The Story of the Mushets* (London: Nelson, 1952).

Phillips, P., *The Forth Bridge* (Edinburgh: Grant, 1899).

8. NEW LAMPS FOR OLD

Belliss, J. Edward, 'A History of G.E. Belliss & Company and Belliss & Morcom, Ltd', *Newcomen Society Transactions*, Vol. XXXVII, 1964–5.

Crompton, R.E.B., *Reminiscences* (London: Constable, 1928).

Durham, John, *Telegraphs in Victorian London* (Cambridge: The Golden Head Press, 1959).

Klapper, Charles, *The Golden Age of Tramways* (London: Routledge & Kegan Paul, 1961).

Ridding, Arthur, *S.Z. de Ferranti, Pioneer of Electric Power*, A Science Museum Booklet (London: H.M.S.O., 1964).

Willans, K.W., 'Peter William Willans', *Newcomen Society Transactions*, Vol. XVIII, 1951–3.

Williams, L. Pearce, *Michael Faraday* (London: Chapman & Hall, 1965).

9. CIVIL ENGINEERING AFTER 1860

Barker, T.C. and Robbins, Michael, *A History of London Transport, Vol. I, The Nineteenth Century* (London: Allen & Unwin, 1962).

Broodbank, Sir Joseph G., *History of the Port of London* (London: Daniel O'Connor, 1921).

Jackson, Alan A. and Croome, Desmond, *Rails Through the Clay, A History of London's Tube Railways* (London: Allen & Unwin, 1962).

Mackay, J., *Life of Sir John Fowler* (London: John Murray, 1900).

'Manchester Ship Canal, The', *Engineering*, 26 January 1894.

Vernon-Harcourt, L.F., *Rivers and Canals*, Vol. II (Oxford: The Clarendon Press, 1896). (The Manchester Ship Canal).

Walker, Thomas A., *The Severn Tunnel, Its Construction and Difficulties* (London: Richard Bentley, 1891).

10 THE SHAPE OF THINGS TO COME

Bird, Anthony, *The Motor Car, 1765–1914* (London: Batsford, 1963).

Gibbs-Smith, Charles A., *The Invention of the Aeroplane, 1799–1909* (London: Faber & Faber, 1966).

King, G.A., '*How the Oil Engine Became Efficient*', Engineering Heritage (London: Heinemann, 1963).

Kingsford, P.W., *F.W. Lanchester, The Life of an Engineer* (London: Arnold, 1960).

Lanchester, G.H., 'F.W. Lanchester, L.L.D., F.R.S., His Life and Work', *Newcomen Society Transactions*, Vol. XXX, 1957.

Index